Recent Advances in Novel Materials for Future Spintronics

Recent Advances in Novel Materials for Future Spintronics

Special Issue Editors

Xiaotian Wang
Hong Chen
Rabah Khenata

MDPI • Basel • Beijing • Wuhan • Barcelona • Belgrade

MDPI

Special Issue Editors

Xiaotian Wang
Southwest University
China

Hong Chen
Southwest University
China

Rabah Khenata
Université de Mascara
Algeria

Editorial Office
MDPI
St. Alban-Anlage 66
4052 Basel, Switzerland

This is a reprint of articles from the Special Issue published online in the open access journal *Applied Sciences* (ISSN 2076-3417) from 2018 to 2019 (available at: https://www.mdpi.com/journal/applsci/special_issues/materials_spintronics).

For citation purposes, cite each article independently as indicated on the article page online and as indicated below:

LastName, A.A.; LastName, B.B.; LastName, C.C. Article Title. *Journal Name* **Year**, *Article Number*, Page Range.

ISBN 978-3-03897-976-0 (Pbk)
ISBN 978-3-03897-977-7 (PDF)

Contents

About the Special Issue Editors

Xiaotian Wang, Ph.D., is an associate professor at the School of Physical Science and Technology, Southwest University in China. His research interests primarily include the following: total energy Density functional theory + on-site Coulomb interaction (DFT+U) or Density functional theory + dynamical mean-field theory (DFT+DMFT) calculations of structural, electronic, mechanical, and thermodynamic properties and high-pressure behavior of strongly correlated electronic systems; DFT studies of topological insulators; and DFT studies of half-metallic materials and spin-gapless semiconductors.

Hong Chen, Ph.D., obtained his Ph.D. degree in particle physics from the Institute of High Energy Physics of the Chinese Academy of Sciences (CAS) in 1997. Currently, he is a professor at the School of Physical Science and Technology, Southwest University (SWU) in China. He is also the scientific supervisor of Institute of Modern Physics of SWU. His research interests primarily include the following: theoretical studies of hadronic and nuclear structures; and first-principles studies of solid materials and clusters.

Rabah Khenata, Ph.D., received his Ph.D. from Sidi Bel Abbes University in 2005. Currently, he is the head of the Laboratoire de Physique Quantique, de la Matière et de la Modélisation Mathématique (LPQ3M) at Mascara University. He is a professor of computational physics and a founding member of the Academy of Science and Technology in Algeria. His main scientific work is focused on the structural, mechanical, magnetic, and optoelectronic properties of crystalline materials using density functional theory (DFT) as implemented in some computer packages.

Preface to "Recent Advances in Novel Materials for Future Spintronics"

Spintronics, which uses the spins of electrons as information carriers and possesses the potential advantages of speeding up data processing, high circuit integration density, and low energy consumption, can be seen as one of the most promising next-generation information technologies. To date, it must be noted that spintronics has faced a number of challenges limiting its widespread use, including spin generation and injection, long-distance spin transport, and manipulation and detection of spin orientation. To solve these issues, many new concepts and spintronics materials have been proposed, such as half-metals, spin-gapless semiconductors, and bipolar magnetic semiconductors. In designing these spintronics materials, first-principles calculations play a very important role. This book is based on the Special Issue of the journal *Applied Sciences* on 'Recent Advances in Novel Materials for Future Spintronics'. This collection of first-principles research articles includes topics such as recent advances in newly predicted half-metallic materials, new attempts in strain tuneable quaternary spintronic Heusler compounds, recent progress in surface and device investigations based on bulk-type spin-gapless semiconductors, frontiers in skyrmionic phase behavior of novel films, and potential for furthering spintronic materials development.

Xiaotian Wang, Hong Chen, Rabah Khenata
Special Issue Editors

applied
sciences

MDPI

Editorial

Special Issue on "Recent Advances in Novel Materials for Future Spintronics"

Xiaotian Wang [1,*], **Rabah Khenata** [2,*] **and Hong Chen** [1,*]

1 School of Physical Science and Technology, Southwest University, Chongqing 400715, China
2 Laboratoire de Physique Quantique de la Matière et de Modélisation Mathématique (LPQ3M), Université de Mascara, Mascara 29000, Algeria
* Correspondence: xiaotianwang@swu.edu.cn (X.W.); khenata_rabah@yahoo.fr (R.K.); chenh@swu.edu.cn (H.C.)

Received: 18 April 2019; Accepted: 26 April 2019; Published: 28 April 2019

1. Referees for the Special Issue

A total of 23 manuscripts were received for our Special Issue (SI), of which 7 manuscripts were directly rejected without peer review. The remaining 16 articles were all strictly reviewed by no less than two reviewers in related fields. Finally, 13 of the manuscripts were recommended for acceptance and published in *Applied Sciences-Basel*. Referees from 10 different countries provided valuable suggestions for the manuscripts in our SI, the top five being the USA, Germany, Korea, Spain, and Finland. The names of these distinguished reviewers are listed in Table A1. We would like to thank all of these reviewers for their time and effort in reviewing the papers in our SI.

2. Main Content of the Special Issue

Since tetragonal Heusler compounds have many potential applications in spintronics and magnetoelectric devices, such as ultrahigh-density spintronic devices, spin transfer torque devices, and permanent magnets, they have received extensive attention in recent years [1–5]. In this SI, Zhang et al. [6] studied the magnetic and electronic structures of cubic and tetragonal types of Mn_3Z (Z = Al, Ga, In, Tl, Ge, Sn, Pb) Heusler alloys. The authors used first-principles calculations to describe the impact of increasing atomic radius on the structure and properties of Heusler alloys. They investigated tetragonal distortions in relation to different volumes for Mn_3Ga alloys and extended this analysis to other elements by replacing Ga with Al, In, Tl, Si, Ge, Sn, and Pb.

Spintronics has many advantages over traditional electronics, such as no volatility, high data processing speed, low energy consumption, and high integration density. Therefore, spintronics, which utilizes spin instead of charge as the carrier for information transportation and processing, can be seen as one of the most promising ways to implement high-speed and low-energy electronic devices. However, in the process of developing spintronic devices, we have also encountered many bottlenecks, including spin-polarized carrier generation and injection, long-range spin-polarization transport, and spin manipulation and detection. To overcome these problems, various types of spintronic materials have been proposed, such as spin-gapless semiconductors (SGSs) [7–13], Dirac half-metals [14,15], diluted magnetic semiconductors (DMSs) [16,17], and bipolar magnetic semiconductors (BMSs) [18–20]. In this SI, Liu et al. [21] predicted two new 1:1:1:1 quaternary Heusler alloys, ZrRhTiAl and ZrRhTiGa, and studied their mechanical, magnetic, electronic, and half-metallic properties via first principles. Chen et al. [22] investigated the effect of main-group element doping on the magnetism, half-metallic property, Slater–Pauling rule, and electronic structures of the TiZrCoIn alloy. Feng et al. [23] calculated the band structures, density of states, magnetic moments, and the band-gap of two quaternary Heusler half-metals, FeRhCrSi and FePdCrSi, by means of first principles. Zhang et al. [24] performed first-principles calculation to investigate the electronic structure of half-metallic Prussian blue analogue

GaFe(CN)$_6$. They revealed its magnetic and mechanical properties. The pressure dependence of the electronic structure was also investigated in their study. In 2017, Wang et al. [25] predicted a rare strain-tunable electronic band structure, which can be utilized in spintronics. Based on Wang et al.'s study, Chen et al. [26] demonstrated that the physical state of ScFeRhP can be tuned by uniform strain. Theoretical predictions of strain-adjustable quaternary spintronic Heusler compounds remain of high importance in the field of spintronics. Similar works can also be found in References [27–32].

In recent years, SGSs [33] have attracted widespread attention in the field of spintronics. Thus far, nearly 100 Heusler-type SGSs have been theoretically predicted, of which Mn$_2$CoAl, Ti$_2$CoAl, and Ti$_2$CoSi have been extensively studied. In this SI, Wei, Wu, and Feng et al. focused on these novel materials. Wei et al. [34] studied the interfacial electronic, magnetic, and spin transport properties of Mn$_2$CoAl/Ag/Mn$_2$CoAl current-perpendicular-to-plane spin valves (CPP-SV) based on density functional theory and non-equilibrium Green's function. Wu et al. [35] conducted a comprehensive study of the electronic and magnetic properties of the Ti$_2$CoAl/MgO (100) heterojunction with first-principles calculations. Ten potential Ti$_2$CoAl/MgO (100) junctions are presented based on the contact between the possible atomic interfaces. The atom-resolved magnetic moments at the interface and subinterface layers were calculated and compared with the values obtained from bulk materials. The spin polarizations were calculated to further illustrate the effective range of tunnel magnetoresistance (TMR) values. Feng et al. [36] systematically investigated the effect of Fe doping in Ti$_2$CoSi and observed the transition from gapless semiconductor to nonmagnetic semiconductor.

Chen et al. [37] used the spin-polarized density functional theory based on first-principles methods to investigate the electronic and magnetic properties of bulk and monolayer CrSi$_2$. Their calculations show that the bulk form of CrSi$_2$ is a nonmagnetic semiconductor with a band gap of 0.376 eV. Interestingly, there are claims that the monolayer of CrSi$_2$ is metallic and ferromagnetic in nature, which is attributed to the quantum size and surface effects of the monolayer.

Jekal et al. [38] conducted a theoretical investigation with the help of the density functional theory and showed that the creation of small, isolated, and stabilized skyrmions with an extremely reduced size of a few nanometers in GdFe$_2$ films can be predicted by 4d and 5d TM (transition metal) capping. Magnetic skyrmions is an exciting area of research and has gained much attention from researchers all over the world. We hope that this work may add value to the scientific community and be helpful for reference in future work.

Finally, we introduce two manuscripts in this SI related to computational materials. Although these two papers are not in the field of spintronics, they belong to the field of computational materials science. The interaction of hydrogen with metal surfaces is an interesting topic in the scientific and engineering world. In this SI, Wu et al. [39] investigated the hydrogen adsorption and diffusion processes on a Mo-doped Nb (100) surface and found that the H atom is stabilized at the hollow sites. They also evaluated the energy barrier along the HS→TIS pathway. Due to their unique physical properties and wide application, Bi-based oxides have received extensive attention in the fields of multiferroics, superconductivity, and photocatalysis. In this SI, Liu et al. [40] investigated the electronic structure as well as the optical, mechanical, and lattice dynamic properties of tetragonal MgBi$_2$O$_6$ using the first-principles method.

Funding: This research was funded by the Program for Basic Research and Frontier Exploration of Chongqing City (Grant No. cstc2018jcyjA0765), the National Natural Science Foundation of China (Grant No. 51801163), and the Doctoral Fund Project of Southwest University, China (Grant No. 117041).

Acknowledgments: We would like to sincerely thank our assistant editor, Emily Zhang (emily.zhang@mdpi.com), for all the efforts she has made for this Special Issue in the past few months.

Conflicts of Interest: The authors declare no conflict of interest.

Appendix A

Table A1. SI reviewer list.

Antonio Frontera	Attila Kákay	Anton O. Oliynyk	Akinola Oyedele
Bhagwati Prasad	David L. Huber	Élio Alberto Périgo	Guangming Cheng
Hannes Rijckaert	Jae Hoon Jang	Jesús López-Sánchez	Ji-Sang Park
Kaupo Kukli	Lalita Saharan	Marijan Beg	Michael Leitner
Masayuki Ochi	Ning Kang	Norbert M. Nemes	Supriyo Bandyopadhyay
Shuo Chen	Soumyajyoti Haldar	Suranjan Shil	Torbjörn Björkman
Uwe Stuhr	Weon Ho Shin	Xueqiang Alex Zhang	Masayuki Ochi
Byeongchan Lee			

References

1. Nayak, A.K.; Shekhar, C.; Winterlik, J.; Gupta, A.; Felser, C. Mn$_2$PtIn: A tetragonal Heusler compound with exchange bias behavior. *Appl. Phys. Lett.* **2012**, *100*, 152404. [CrossRef]
2. Faleev, S.V.; Ferrante, Y.; Jeong, J.; Samant, M.G.; Jones, B.; Parkin, S.S.P. Origin of the tetragonal ground state of Heusler compounds. *Phys. Rev. Appl.* **2017**, *7*, 034022. [CrossRef]
3. Liu, Z.H.; Tang, Z.; Tan, J.G.; Zhang, Y.J.; Wu, Z.G.; Wang, X.T.; Liu, G.D.; Ma, X.Q. Tailoring structural and magnetic properties of Mn3− xFexGa alloys towards multifunctional applications. *IUCrJ* **2018**, *5*, 794–800. [CrossRef]
4. Faleev, S.V.; Ferrante, Y.; Jeong, J.; Samant, M.G.; Jones, B.; Parkin, S.S.P. Heusler compounds with perpendicular magnetic anisotropy and large tunneling magnetoresistance. *Phys. Rev. Mater.* **2017**, *1*, 024402. [CrossRef]
5. Wu, M.; Han, Y.; Bouhemadou, A.; Cheng, Z.; Khenata, R.; Kuang, M.; Wang, X.; Yang, T.; Yuan, H.; Wang, X. Site preference and tetragonal distortion in palladium-rich Heusler alloys. *IUCrJ* **2019**, *6*, 218–225. [CrossRef] [PubMed]
6. Zhang, H.; Liu, W.; Lin, T.; Wang, W.; Liu, G. Phase Stability and Magnetic Properties of Mn$_3$Z (Z = Al, Ga, In, Tl, Ge, Sn, Pb) Heusler Alloys. *Appl. Sci.* **2019**, *9*, 964. [CrossRef]
7. Gao, Q.; Opahle, I.; Zhang, H. High-throughput screening for spin-gapless semiconductors in quaternary Heusler compounds. *Phys. Rev. Mater.* **2019**, *3*, 024410. [CrossRef]
8. Wang, X.; Li, T.; Cheng, Z.; Wang, X.L.; Chen, H. Recent advances in Dirac spin-gapless semiconductors. *Appl. Phys. Rev.* **2018**, *5*, 041103. [CrossRef]
9. Han, Y.; Khenata, R.; Li, T.; Wang, L.; Wang, X. Search for a new member of parabolic-like spin-gapless semiconductors: The case of diamond-like quaternary compound CuMn$_2$InSe$_4$. *Results Phys.* **2018**, *10*, 301–303. [CrossRef]
10. Venkateswara, Y.; Gupta, S.; Samatham, S.S.; Varma, M.R.; Suresh, K.G.; Alam, A. Competing magnetic and spin-gapless semiconducting behavior in fully compensated ferrimagnetic CrVTiAl: Theory and experiment. *Phys. Rev. B* **2018**, *97*, 054407. [CrossRef]
11. Wang, X.L. Proposal for a new class of materials: Spin gapless semiconductors. *Phys. Rev. Lett.* **2008**, *100*, 156404. [CrossRef] [PubMed]
12. Tas, M.; Şaşıoğlu, E.; Friedrich, C.; Galanakis, I. A first-principles DFT+ GW study of spin-filter and spin-gapless semiconducting Heusler compounds. *J. Magn. Magn. Mater.* **2017**, *441*, 333–338. [CrossRef]
13. Liu, Y.; Bose, S.K.; Kudrnovský, J. 4-d magnetism: Electronic structure and magnetism of some Mo-based alloys. *J. Magn. Magn. Mater.* **2017**, *423*, 12–19. [CrossRef]
14. Jiao, Y.; Ma, F.; Zhang, C.; Bell, J.; Sanvito, S.; Du, A. First-principles prediction of spin-polarized multiple Dirac rings in manganese fluoride. *Phys. Rev. Lett.* **2017**, *119*, 016403. [CrossRef] [PubMed]
15. Ma, F.; Jiao, Y.; Jiang, Z.; Du, A. Rhombohedral Lanthanum Manganite: A New Class of Dirac Half-Metal with Promising Potential in Spintronics. *ACS Appl. Mater. Interfaces* **2018**, *10*, 36088–36093. [CrossRef] [PubMed]

16. Goumrhar, F.; Bahmad, L.; Mounkachi, O.; Benyoussef, A. Magnetic properties of vanadium doped CdTe: Ab initio calculations. *J. Magn. Magn. Mater.* **2017**, *428*, 368–371. [CrossRef]

17. Pereira, L.M.C. Experimentally evaluating the origin of dilute magnetism in nanomaterials. *J. Phys. D: Appl. Phys.* **2017**, *50*, 393002. [CrossRef]

18. Farghadan, R. Bipolar magnetic semiconductor in silicene nanoribbons. *J. Magn. Magn. Mater.* **2017**, *435*, 206–211. [CrossRef]

19. Zha, X.H.; Ren, J.C.; Feng, L.; Bai, X.; Luo, K.; Zhang, Y.; He, J.; Huang, Q.; Francisco, J.S.; Du, S. Bipolar magnetic semiconductors among intermediate states during the conversion from $Sc_2C(OH)_2$ to Sc_2CO_2 MXene. *Nanoscale* **2018**, *10*, 8763–8771. [CrossRef]

20. Cheng, H.; Zhou, J.; Yang, M.; Shen, L.; Linghu, J.; Wu, Q.; Qian, P.; Feng, Y.P. Robust two-dimensional bipolar magnetic semiconductors by defect engineering. *J. Mater. Chem. C* **2018**, *6*, 8435–8443. [CrossRef]

21. Liu, W.; Zhang, X.; Jia, H.; Khenata, R.; Dai, X.; Liu, G. Theoretical Investigations on the Mechanical, Magneto-Electronic Properties and Half-Metallic Characteristics of ZrRhTiZ (Z = Al, Ga) Quaternary Heusler Compounds. *Appl. Sci.* **2019**, *9*, 883. [CrossRef]

22. Chen, Y.; Chen, S.; Wang, B.; Wu, B.; Huang, H.; Qin, X.; Li, D.; Yan, W. Half-Metallicity and Magnetism of the Quaternary Heusler Compound $TiZrCoIn_{1-x}Ge_x$ from the First-Principles Calculations. *Appl. Sci.* **2019**, *9*, 620. [CrossRef]

23. Feng, L.; Ma, J.; Yang, Y.; Lin, T.; Wang, L. The Electronic, Magnetic, Half-Metallic and Mechanical Properties of the Equiatomic Quaternary Heusler Compounds FeRhCrSi and FePdCrSi: A First-Principles Study. *Appl. Sci.* **2018**, *8*, 2370. [CrossRef]

24. Zhang, C.; Huang, H.; Luo, S. First Principles Study on the Effect of Pressure on the Structure, Elasticity and Magnetic Properties of Cubic $GaFe(CN)_6$ Prussian Blue Analogue. *Appl. Sci.* **2019**, *9*, 1607. [CrossRef]

25. Wang, X.; Cheng, Z.; Liu, G.; Dai, X.; Khenata, R.; Wang, L.; Bouhemadou, A. Rare earth-based quaternary Heusler compounds MCoVZ (M = Lu, Y; Z = Si, Ge) with tunable band characteristics for potential spintronic applications. *IUCrJ* **2017**, *4*, 758–768. [CrossRef]

26. Chen, Z.; Rozale, H.; Gao, Y.; Xu, H. Strain Control of the Tunable Physical Nature of a Newly Designed Quaternary Spintronic Heusler Compound ScFeRhP. *Appl. Sci.* **2018**, *8*, 1581. [CrossRef]

27. Zhu, S.; Li, T. Strain-induced programmable half-metal and spin-gapless semiconductor in an edge-doped boron nitride nanoribbon. *Phys. Rev. B* **2016**, *93*, 115401. [CrossRef]

28. Gao, G.; Ding, G.; Li, J.; Yao, K.; Wu, M.; Qian, M. Monolayer MXenes: promising half-metals and spin gapless semiconductors. *Nanoscale* **2016**, *8*, 8986–8994. [CrossRef]

29. Wang, X.; Cheng, Z.; Khenata, R.; Wu, Y.; Wang, L.; Liu, G. Lattice constant changes leading to significant changes of the spin-gapless features and physical nature in a inverse heusler compound Zr_2MnGa. *J. Magn. Magn. Mater.* **2017**, *444*, 313–318. [CrossRef]

30. Wang, X.; Cheng, Z.; Khenata, R.; Rozale, H.; Wang, J.; Wang, L.; Guo, R.; Liu, G. A first-principle investigation of spin-gapless semiconductivity, half-metallicity, and fully-compensated ferrimagnetism property in Mn_2ZnMg inverse Heusler compound. *J. Magn. Magn. Mater.* **2017**, *423*, 285–290. [CrossRef]

31. Zhang, Y.J.; Liu, Z.H.; Liu, E.K.; Liu, G.D.; Ma, X.Q.; Wu, G.H. Towards fully compensated ferrimagnetic spin gapless semiconductors for spintronic applications. *EPL* **2015**, *111*, 37009. [CrossRef]

32. Wang, X.T.; Cheng, Z.X.; Wang, J.L.; Rozale, H.; Wang, L.Y.; Yu, Z.Y.; Yang, J.T.; Liu, G.D. Strain-induced diverse transitions in physical nature in the newly designed inverse Heusler alloy Zr_2MnAl. *J. Alloys Compd.* **2016**, *686*, 549–555. [CrossRef]

33. Wang, X.; Cheng, Z.; Wang, J.; Wang, X.L.; Liu, G. Recent advances in the Heusler based spin-gapless semiconductors. *J. Mater. Chem. C* **2016**, *4*, 7176–7192. [CrossRef]

34. Wei, M.-S.; Cui, Z.; Ruan, X.; Zhou, Q.-W.; Fu, X.-Y.; Liu, Z.-Y.; Ma, Q.-Y.; Feng, Y. Interface Characterization of Current-Perpendicular-to-Plane Spin Valves Based on Spin Gapless Semiconductor Mn_2CoAl. *Appl. Sci.* **2018**, *8*, 1348. [CrossRef]

35. Wu, B.; Huang, H.; Zhou, G.; Feng, Y.; Chen, Y.; Wang, X. Structure, Magnetism, and Electronic Properties of Inverse Heusler Alloy $Ti_2CoAl/MgO(100)$ Herterojuction: The Role of Interfaces. *Appl. Sci.* **2018**, *8*, 2336. [CrossRef]

36. Feng, Y.; Cui, Z.; Wei, M.-S.; Wu, B.; Azam, S. Spin Gapless Semiconductor–Nonmagnetic Semiconductor Transitions in Fe-Doped Ti_2CoSi: First-Principle Calculations. *Appl. Sci.* **2018**, *8*, 2200. [CrossRef]

37.   Chen, S.; Chen, Y.; Yan, W.; Zhou, S.; Qin, X.; Xiong, W.; Liu, L. Electronic and Magnetic Properties of Bulk and Monolayer CrSi$_2$: A First-Principle Study. *Appl. Sci.* **2018**, *8*, 1885. [CrossRef]

38.   Jekal, S.; Danilo, A.; Phuong, D.; Zheng, X. First-Principles Prediction of Skyrmionic Phase Behavior in GdFe$_2$ Films Capped by 4d and 5d Transition Metals. *Appl. Sci.* **2019**, *9*, 630. [CrossRef]

39.   Wu, Y.; Wang, Z.; Wang, D.; Qin, J.; Wan, Z.; Zhong, Y.; Hu, C.; Zhou, H. First-Principles Investigation of Atomic Hydrogen Adsorption and Diffusion on/into Mo-doped Nb (100) Surface. *Appl. Sci.* **2018**, *8*, 2466. [CrossRef]

40.   Liu, L.; Wang, D.; Zhong, Y.; Hu, C. Electronic, Optical, Mechanical and Lattice Dynamical Properties of MgBi$_2$O$_6$: A First-Principles Study. *Appl. Sci.* **2019**, *9*, 1267. [CrossRef]

applied
sciences

MDPI

Article

First Principles Study on the Effect of Pressure on the Structure, Elasticity, and Magnetic Properties of Cubic GaFe(CN)$_6$ Prussian Blue Analogue

Chuankun Zhang, Haiming Huang * and **Shijun Luo**

School of Science, Hubei University of Automotive Technology, Shiyan 442002, China;
zhangchk_lx@huat.edu.cn (C.Z.); luosjhuat@163.com (S.L.)
* Correspondence: smilehhm@163.com

Received: 17 March 2019; Accepted: 16 April 2019; Published: 18 April 2019

Abstract: The structure, elasticity, and magnetic properties of Prussian blue analogue GaFe(CN)$_6$ under external pressure ranges from 0 to 40 GPa were studied by first principles calculations. In the range of pressure from 0 to 35 GPa, GaFe(CN)$_6$ not only has the half-metallic characteristics of 100% spin polarization, but also has stable mechanical properties. The external pressure has no obvious effect on the crystal structure and anisotropy of GaFe(CN)$_6$, but when the pressure exceeds 35 GPa, the half-metallicity of GaFe(CN)$_6$ disappears, the mechanical properties are no longer stable, and total magnetic moments per formula unit are no longer integer values.

Keywords: half-metallic material; first principles; Prussian blue analogue; pressure

1. Introduction

Whether spin-polarized electrons can be efficiently injected into semiconductor materials is one of the key technologies to realize spintronic devices [1–6]. Previous studies have shown that magnetic materials with high spin polarizability can effectively inject spin-polarized electrons [7–10]. Half-metallic ferromagnets with a high Curie temperature and nearly 100% spin polarizability undoubtedly become the most ideal spin electron injection source for semiconductors. Among the two different spin channels of half-metallic ferromagnets, one spin channel is metallic, while the other is insulating or a semiconductor [11]. Half-metallic ferromagnets are widely used in spin diodes, spin valves, and spin filters because of their unique electronic structure [12–15].

Since the first half-metallic ferromagnet was predicted by theory, after more than 30 years of development, half-metallic ferromagnetic materials have become a hot topic in materials science and condensed matter physics. Up to now, half-metallic ferromagnets have been found mainly as follows: ternary metal compounds represented by Heulser alloy [16–19], magnetic metal oxides [20,21], perovskite compounds [22,23], dilute magnetic semiconductors [24,25], zinc-blende type pnictides and chalcogenides [26,27], organic–inorganic hybrid compounds [28,29]. Even some two-dimensional materials have half-metallic ferromagnets [30–33].

Prussian blue analogs are a class of metal-organic frameworks with a simple cubic structure, whose chemical formula can be expressed as A$_2$M[M(CN)$_6$] (A = alkaline metal ions, zeolitic water; M/M'= Fe, Co, Mn, etc.) [34]. Prussian blue analogs often have simpler molecular configurations due to the existence of vacancy defects. In Prussian blue analogs, there is a large space between metal ions and -CN- groups, which can effectively accommodate alkali metal ions such as Li$^+$, Na$^+$, and K$^+$. The open structure of Prussian blue analogs makes it exhibit excellent electrochemical performance [35–37].

The magnetic study of Prussian blue analogs has also attracted people's attention for a long time. In 1999, Holmes et al. reported a compound KV[Cr(CN)$_6$] with a Curie temperature as high as 376 K [38]. In 2003, Sato et al. proposed that electrochemical methods could be used to control the magnetism and Curie temperature of Prussian blue analogs [39]. They also pointed out that it was feasible and promising to control the magnetism of Prussian blue analogs by light. Half-metals have also been found in these compounds by studying the magnetism. Two well-defined Prussian blue analogues are predicted as half-metallicity using first principles [40]. In the present study, we will study the structure, elasticity, and magnetic properties of a new Prussian blue analogue GaFe(CN)$_6$ under pressure and predict that the compound is half-metallic.

2. Materials and Methods

The projector augmented wave (PAW) [41] method encoded in the software Vienna Ab initio Simulation Package (VASP) [42] was performed during the calculations. The generalized gradient approximation (GGA) of the Perdew–Burke–Ernzerhof (PBE) functional is used as exchange correlation potential [43]. The electronic configurations—$4s^24p^1$ for Ga, $4s^23d^6$ for Fe, $2s^22p^2$ for C, and $2s^22p^3$ for N—were treated as valence electrons in calculations. For the self-consistent calculation, the plane wave cutoff energy was chosen to be 400 eV. A mesh of $9 \times 9 \times 9$ Monkhorst–Pack k-point was used. The convergence tolerances were selected as the difference in total energy and the maximum force within 1.0×10^{-5} eV and 1.0×10^{-2} eV/atom, respectively.

3. Results and Discussion

Crystal structure characterization based on high resolution synchrotron radiation X-ray diffraction shows that the Prussian blue analogue of GaFe(CN)$_6$ is a cubic crystal with space group $Fm\bar{3}m$, as shown in Figure 1. The structure of GaFe(CN)$_6$ is formed with FeC$_6$ and GaN$_6$ octahedrons, which are equivalent to ABX$_3$ type perovskite with vacancy in A site. In the structure of GaFe(CN)$_6$, the -Ga-N≡C-Fe- chain is formed between gallium, carbon, nitrogen, and iron atoms. Experimentally, the lattice constant of GaFe(CN)$_6$ was measured as 10.0641 Å at 273 K [36], and the occupied positions of each atom in the structure are shown in Table 1.

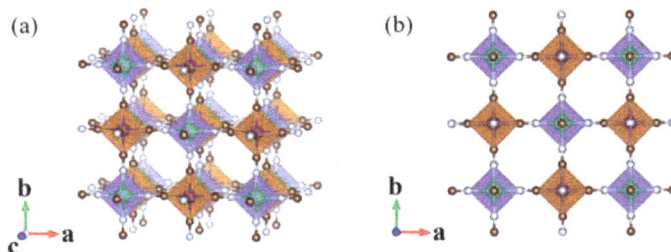

Figure 1. Crystal structure of GaFe(CN)$_6$. (**a**) Side view; (**b**) top view.

Table 1. Atomic occupied positions in GaFe(CN)$_6$.

Atom	Exp.			Present		
	x	y	z	x	y	z
Ga	0.0	0.0	0.0	0.0	0.0	0.0
Fe	0.5	0.0	0.0	0.5	0.0	0.0
C	0.3043	0.0	0.0	0.3253	0.0	0.0
N	0.1883	0.0	0.0	0.2114	0.0	0.0

In order to obtain the theoretical equilibrium lattice constant and the ground state properties of GaFe(CN)$_6$, we constructed supercells based on experimental structural parameters and calculated the total energy of ferromagnetic (FM), non-magnetic (NM), and antiferromagnetic (AFM) states of GaFe(CN)$_6$ under different lattice constants. The ground state is determined based on the principle that the lower the energy is, the more stable the structure is. The total energies of GaFe(CN)$_6$ in

FM, NM, and AFM states are drawn in Figure 2. Obviously, FM states have lower total energy than NM and AFM states, which means the ferromagnetic state is the most stable for $GaFe(CN)_6$. The equilibrium lattice constant obtained at the same time was 10.1883 Å. This result is slightly larger than the experimental result, and the deviation is 1.23% compared with the experimental result, which is within a reasonable range. The coordinates of the positions of the atoms in the equilibrium state of $GaFe(CN)_6$ are also listed in Table 1. Excepting that the x coordinates of C and N atoms deviate from the experimental data, the other results are consistent with the experimental values.

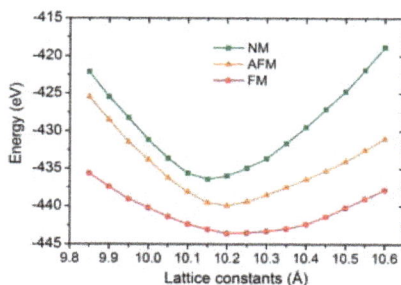

Figure 2. The total energies of $GaFe(CN)_6$ in ferromagnetic (FM), non-magnetic (NM), and antiferromagnetic (AFM) states.

In order to study the effect of pressure on the crystal structure of $GaFe(CN)_6$, the pressure measurement of $GaFe(CN)_6$ was carried out at intervals of 5.0 GPa under pressure of 0–40 GPa. The variation of relative lattice constant a/a_0 and relative volume V/V_0 with pressure was obtained, as shown in Figure 3. Among them, a_0 is the equilibrium lattice constant at 0 GPa and V_0 is the cell volume at 0 GPa. As can be seen from Figure 3, the lattice constant decreases gradually with the increase of external pressure, resulting in the corresponding decrease of volume V and relative volume V/V_0.

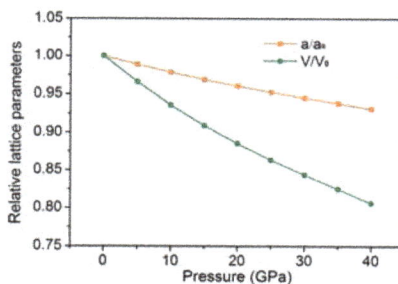

Figure 3. The variation of relative lattice constant a/a_0 and relative volume V/V_0 with pressure.

In order to further understand the variation of structural parameters with pressure, the curve of Figure 3 is fitted and calculated, and the binary quadratic state equations of a/a_0 and V/V_0 of $GaFe(CN)_6$ and pressure are obtained, as shown below.

$$a/a_0 = 0.99645 - 0.00171P + 5.71387 \times 10^{-5}P^2 \tag{1}$$

$$V/V_0 = 0.98777 - 0.00475P + 4.05769 \times 10^{-4}P^2 \tag{2}$$

Table 2 gives the structural parameters of $GaFe(CN)_6$ under pressure. The lattice constant at 40 GPa is 9.4828 Å, which is only 93.1% of the lattice constant at 0 GPa. The bond lengths of C–N,

Ga–N, and Fe–C in the compounds decrease with the increase of pressure, which is mainly due to the compression of the volume of the compounds under pressure and the reduction of the spacing between atoms. The pressure from 0 to 40 GPa does not cause structural transition of GaFe(CN)$_6$, because GaFe(CN)$_6$ still presents a cubic phase structure. Except for the x-direction coordinates of C and N atoms, the positions or coordinates of other atoms in compounds have not changed.

Table 2. Structural parameters of GaFe(CN)$_6$ under different pressures.

Pressure	a (Å)	C–N(Å)	Ga–N(Å)	Fe–C(Å)	C(x,0,0)	N(x,0,0)
0	10.1883	1.160	2.155	1.780	0.32533	0.21148
5	10.0706	1.156	2.118	1.762	0.32508	0.21029
10	9.9649	1.152	2.085	1.745	0.32492	0.20928
15	9.8695	1.149	2.057	1.729	0.32481	0.20843
20	9.7830	1.145	2.028	1.719	0.32430	0.20728
25	9.7015	1.142	2.008	1.701	0.32471	0.20701
30	9.6271	1.138	1.987	1.688	0.32461	0.20636
35	9.5563	1.135	1.967	1.676	0.32459	0.20579
40	9.4828	1.132	1.945	1.665	0.32447	0.20512

The elastic constants are important parameters reflecting the mechanical stability of the compounds [44,45]. At 0 GPa, the elastic constants C_{11}, C_{12}, and C_{44} of GaFe(CN)$_6$ are 206.7, 53.2, and 54.6 GPa, respectively. The mechanical stability Born–Huang criteria of cubic crystal are expressed as [46,47]:

$$C_{11} - C_{12} > 0, C_{11} + 2C_{12} > 0, C_{44} > 0. \tag{3}$$

The elastic constants of GaFe(CN)$_6$ at 0 GPa satisfy the above conditions, which means that GaFe(CN)$_6$ has stable mechanical properties in an equilibrium state. At the same time, it was noted that the unidirectional elastic constant C_{11} is higher than C_{44}, which indicates that GaFe(CN)$_6$ has weaker resistance to the pure shear deformation compared to the resistance of the unidirectional compression.

Some mechanical parameters can be calculated by elastic constants according to some formulas, which can be obtained in our previous studies [48]. The elastic anisotropy factor A is calculated by the following formula:

$$A = 2C_{44}/(C_{11} - C_{12}). \tag{4}$$

The elastic anisotropy factor A of GaFe(CN)$_6$ is 0.71; it is usually used to quantify the elastic anisotropy and the degree of elastic anisotropy of the compound. In general, the elastic anisotropic factor for isotropic crystals is A = 1, while for anisotropic crystals A ≠ 1. According to this criterion, GaFe(CN)$_6$ is an anisotropic compound. The Poisson's ratio, which reflects the binding force characteristics, is often between 0.25 and 0.50. The Poisson's ratio of GaFe(CN)$_6$ is 0.25, which is just in the range of values, meaning that the inter-atomic forces are central for the compounds. The Debye temperature of the GaFe(CN)$_6$ is 738.4 K, which is calculated from a formula in [47,49].

Under the isotropic pressure, the elastic constants are transformed into the corresponding stress–strain coefficients by the following expressions:

$$B_{11} = C_{11} - P, B_{12} = C_{12} + P, B_{44} = C_{44} - P. \tag{5}$$

The mechanical stability of GaFe(CN)$_6$ under isotropic pressure is determined by the following formula [48,50]:

$$B_{11} - B_{12} > 0, B_{11} + 2B_{12} > 0, B_{44} > 0. \tag{6}$$

The P in the formula above refers to the external pressure. The curves of $B_{11} - B_{12}$, $B_{11} + 2B_{12}$, and B_{44} with pressure are plotted in Figure 4. $B_{11} - B_{12}$ and $B_{11} + 2B_{12}$ increase with the increase of pressure, and also meet the mechanical stability criterion under pressure. When the pressure is greater than 35 GPa, the value of B_{44} is negative, and the stability condition of B_{44} is not satisfied. Generally speaking, when the external pressure of GaFe(CN)$_6$ is less than 35 GPa, its mechanical performance is stable. Once the external pressure exceeds 35 GPa, the mechanical performance of GaFe(CN)$_6$ is unstable.

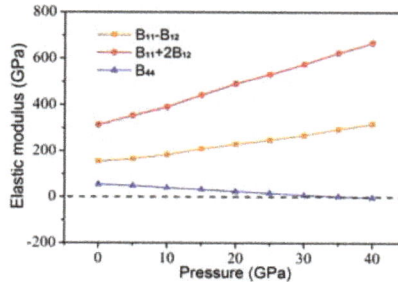

Figure 4. Elastic modulus of GaFe(CN)$_6$ under different pressures.

From 0 to 40 GPa, elastic anisotropy factor A becomes smaller and smaller, and the anisotropic characteristics of GaFe(CN)$_6$ become more obvious. At the same time, the bulk modulus increases from 104.3 to 208.8 GPa, and the Debye temperature reaches 798.5 K. The increase in pressure makes the atoms more closely linked, which makes the compound's stiffness.

The spin-polarized band structures and density of states of GaFe(CN)$_6$ at 0 GPa are depicted in Figure 5. It can be clearly seen that the conduction band minimum (CBM) and valence band maximum (VBM) in majority-spin are located at the same highly symmetric G-point, and a band gap of 4.01 eV is formed between the conduction band and the valence band, indicating that this spin direction has insulator behavior. The bands pass through the Fermi level in minority-spin to exhibit a metallic feature. According to the band theory of quantum solid, GaFe(CN)$_6$ is a half-metal with 100% spin polarization.

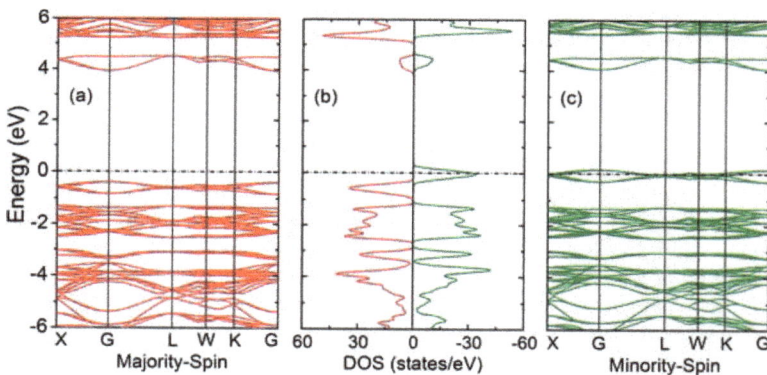

Figure 5. Band structure and density of states (DOS) of GaFe(CN)$_6$ at 0 GPa.

Figure 6 presents the total and local density of state of GaFe(CN)$_6$ at 0 GPa. It can be clearly seen that the half-metallic behavior of GaFe(CN)$_6$ is mainly due to the formation of spin splitting in the vicinity of the Fermi level by the 3d states of the Fe atom and the 2p states of the N atom. The 3d states of the Fe atom and the 2p states of the N atom have obvious spin hybridization in the energy range of −1.01 to 0.35 eV. The 3d state of the Fe atom is also the most important contributor to the total density of GaFe(CN)$_6$. From the magnetic properties generated by spin splitting, it can be inferred that Fe atoms are also the main source of GaFe(CN)$_6$ magnetic moment. In the energy range of −2.7 to −1.01 eV, the density of states is mainly derived from the C-2p, N-2p, and Ga-4P states, and the 3d of the Fe atom has little contribution in this region.

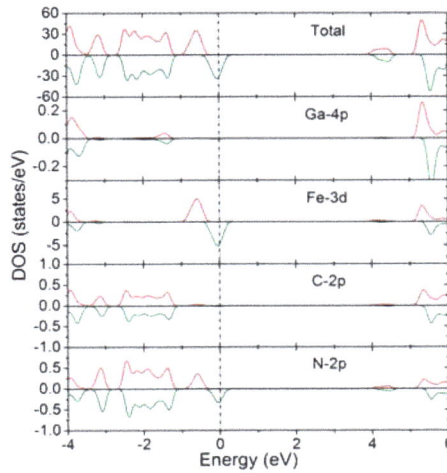

Figure 6. Total and local density of states of $GaFe(CN)_6$ at 0 GPa.

The electronic structure calculation of pressure from 0 to 40 GPa shows that the minority-spin direction of $GaFe(CN)_6$ always shows metallic behavior. In this case, the physical properties of $GaFe(CN)_6$ under pressure are basically determined by the majority-spin electronic states. Figure 7 depicts the CBM and VBM in majority-spin of $GaFe(CN)_6$ as a function of pressure. With the increase of pressure, both CBM and VBM move towards high energy. Once the pressure is greater than 35 GPa, VBM will cross the Fermi level and make $GaFe(CN)_6$ majority-spin also show metallic behavior. In this way, the half-metallicity of $GaFe(CN)_6$ will disappear. It is worth noting that, as can be seen from Figure 7, the density of states across the Fermi level at 40 GPa is very low. This means that the material may not be able to hold enough free electrons and therefore has poor conductivity or metallicity.

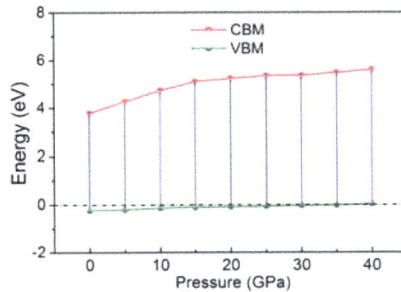

Figure 7. Conduction band minimum (CBM) and valence band maximum (VBM) of $GaFe(CN)_6$ in majority-spin under different pressures.

The effect of pressure on the electronic structure of $GaFe(CN)_6$ can also be confirmed by Figure 8. In Figure 8, we can see that the minority-spin electronic states are hardly affected by external pressures. A slightly more obvious feature is that the conduction band in the high energy region moves toward a higher energy position as the pressure increases. However, this does not change the metallicity of the minority-spin direction. The electronic structure in majority-spin changes are consistent with the analysis in Figure 7. When the pressure is 40 GPa, the valence band in majority-spin crosses the Fermi level.

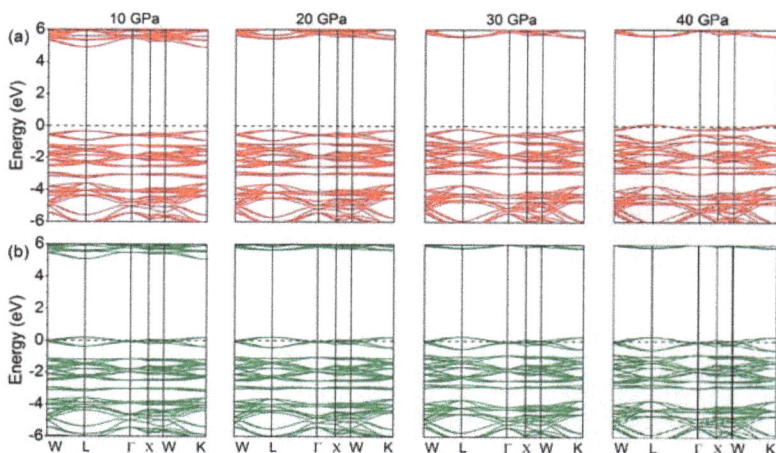

Figure 8. Band structure of GaFe(CN)$_6$ under different pressures. (**a**) Majority-spin; (**b**) minority-spin.

At 0 GPa, the total magnetic moment per formula unit of GaFe(CN)$_6$ is 1.0 μ_B, which is very consistent with the characteristic that the molecular magnetic moment of half-metallic magnetic materials is an integral value. The local magnetic moments of Fe, Ga, C, and N atoms are 0.765 μ_B, −0.007 μ_B, −0.018 μ_B, and 0.035 μ_B, respectively. Obviously, Fe atoms are the most important contributors to the magnetic properties of GaFe(CN)$_6$. The local magnetic moments of Ga, C, and N atoms are very small. Because these three atoms have no magnetism, their magnetic moments are mainly induced by the influence of the Fe atom. In –Ga–N–C–Fe– chemical chains, the local magnetic moments between them show a sign change of −/+/−/+, which means that there is antiferromagnetic coupling between these atoms.

Figure 9 shows the total and local magnetic moments of GaFe(CN)$_6$ under pressure. From 0 to 35 GPa, the total magnetic moment per formula unit of GaFe(CN)$_6$ is 1.0 μ_B. In this pressure range, GaFe(CN)$_6$ has half-metal characteristics. When the pressure exceeds 35 GPa, GaFe(CN)$_6$ is no longer a half-metal and its molecular magnetic moment is no longer an integral value. The local magnetic moment of the Fe atom decreases with the increase of pressure, while the induced magnetic moment of the N atom increases slightly, although its value is very small. The local magnetic moments of Ga and C atoms are hardly affected by pressure. From the local magnetic moment signs of Ga, N, C, and Fe atoms, the change of pressure has no effect on the antiferromagnetic coupling of –Ga–N–C–Fe– chemical chains.

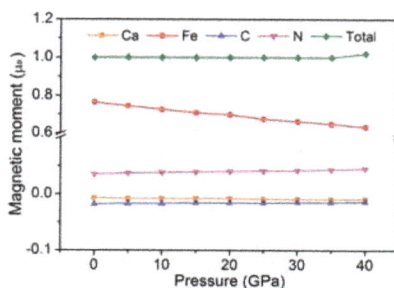

Figure 9. Total and local magnetic moments of GaFe(CN)$_6$ under pressure.

4. Conclusions

First principles calculations were performed to study the structure, elasticity, and magnetism of a Prussian blue analogue GaFe(CN)$_6$ under external pressure ranges from 0 to 40 GPa. The crystal structure obtained by theoretical optimization was very close to the experimental structure, and the external pressure had no obvious effect on the cubic structure of GaFe(CN)$_6$. In the range of pressure from 0 to 35 GPa, GaFe(CN)$_6$ was an anisotropic compound with stable mechanical properties. It also was a half-metallic magnetic material with 100% spin polarization, and its total magnetic moment per formula unit was 1.0 μ_B. When the pressure exceeded 35 GPa, the mechanical properties were no longer stable, the half-metallicity of GaFe(CN)$_6$ disappeared, and the magnetic moment no longer had the typical characteristics of half-metallic magnetic materials, that is, the total magnetic moment per formula unit was no longer an integer value. In terms of magnetism, iron atoms are the most important contributors to GaFe(CN)$_6$ magnetism in the whole pressure range.

Author Contributions: Methodology, C.Z.; investigation, H.H; writing-original draft preparation, H.H.; writing-review and editing, C.Z.; project administration, S.J.

Funding: This work is supported by the Natural Science Foundation of Hubei Province (No. 2017CFB740), the Doctoral Scientific Research Foundation of Hubei University of Automotive Technology (No. BK201804), the Scientific Research Items Foundation of Hubei Educational Committee (No. Q20111801).

Conflicts of Interest: The authors declare no conflict of interest.

References

1. Wolf, S.A.; Awschalom, D.D.; Buhrman, R.A.; Daughton, J.M.; von Molnár, S.; Roukes, M.L.; Chtchelkanova, A.Y.; Treger, D.M. Spintronics: A spin-based electronics vision for the future. *Science* **2001**, *294*, 1488–1495. [CrossRef] [PubMed]
2. Kioseoglou, G.; Hanbicki, A.T.; Sullivan, J.M.; Erve, O.M.J.; Li, C.H.; Erwin, S.C.; Mallory, R.; Yasar, M.; Petrou, A.; Jonker, B.T. Electrical spin injection from an n-type ferromagnetic semiconductor into a III–V device heterostructure. *Nat. Mater.* **2004**, *3*, 799–803. [CrossRef] [PubMed]
3. Jonker, B.T. Progress toward electrical injection of spin-polarized electrons into semiconductors. *IEEE Proc.* **2003**, *91*, 727–740. [CrossRef]
4. Zutic, I.; Fabian, J.; Das Sarma, S. Spintronics: Fundamentals and applications. *Rev. Mod. Phys.* **2004**, *76*, 323–410. [CrossRef]
5. Prinz, G.A. Magnetoelectronics. *Science* **1998**, *282*, 1660–1663. [CrossRef] [PubMed]
6. Flatte, M.E.; Beyers, J.M. Spin diffusion in semiconductors. *Phys. Rev. Lett.* **2000**, *84*, 4220–4223. [CrossRef]
7. Farshchi, R.; Ramsteiner, M. Spin injection from Heusler alloys into semiconductors: A materials perspective. *J. Appl. Phys.* **2013**, *113*, 191101. [CrossRef]
8. Medvedeva, J.E.; Freeman, A.J.; Cui, X.Y.; Stampfl, C.; Newman, N. Half-Metallicity and Efficient Spin Injection in AlN/GaN:Cr (0001) Heterostructure. *Phys. Rev. Lett.* **2005**, *94*, 146602. [CrossRef]
9. De Wijs, G.A.; De Groot, R.A. Towards 100% spin-polarized charge-injection: The half-metallic NiMnSb/CdS interface. *Phys. Rev. B* **2001**, *64*, 020402. [CrossRef]
10. Li, J.; Gao, G.Y.; Min, Y.; Yao, K.L. Half-metallic YN$_2$ monolayer: Dual spin filtering, dual spin diode and spin seebeck effects. *Phys. Chem. Chem. Phys.* **2016**, *18*, 28018–28023. [CrossRef]
11. De Groot, R.A.; Mueller, F.M.; Van Engen, P.G.; Buschow, K.H.J. New Class of Materials: Half-Metallic Ferromagnets. *Phys. Rev. Lett.* **1983**, *50*, 2024. [CrossRef]
12. Han, J.; Shen, J.; Gao, G. CrO$_2$-based heterostructure and magnetic tunnel junction: Perfect spin filtering effect, spin diode effect and high tunnel magnetoresistance. *RSC Adv.* **2019**, *9*, 3550–3557. [CrossRef]
13. Song, Y.; Dai, G. Spin filter and spin valve in ferromagnetic graphene. *Appl. Phys. Lett.* **2015**, *106*, 223104. [CrossRef]
14. Everschor-Sitte, K.; Sitte, M.; MacDonald, A.H. Half-metallic magnetism and the search for better spin valves. *J. Appl. Phys.* **2014**, *116*, 083906. [CrossRef]
15. Stefano, S. Molecular spintronics. *Chem. Soc. Rev.* **2011**, *40*, 3336–3355.

16. Chen, Y.; Chen, S.; Wang, B.; Wu, B.; Huang, H.; Qin, X.; Li, D.; Yan, W. Half-Metallicity and Magnetism of the Quaternary Heusler Compound TiZrCoIn$_{1-x}$Ge$_x$ from the First-Principles Calculations. *Appl. Sci.* **2019**, *9*, 620. [CrossRef]

17. Huang, H.M.; Luo, S.J.; Yao, K.L. Half-metallicity and tetragonal distortion in semi-heusler alloy FeCrSe. *J. Appl. Phys.* **2014**, *115*, 043713. [CrossRef]

18. Wang, X.T.; Dai, X.F.; Wang, L.Y.; Liu, X.F.; Wang, W.H.; Wu, G.H.; Tang, C.C.; Liu, G.D. Electronic structures and magnetism of Rh$_3$Z (Z=Al, Ga, In, Si, Ge, Sn, Pb, Sb) with DO$_3$ structures. *J. Magn. Magn. Mater.* **2015**, *378*, 16–23. [CrossRef]

19. Feng, L.; Ma, J.; Yang, Y.; Lin, T.; Wang, L. The electronic, magnetic, half-metallic and mechanical properties of the equiatomic quaternary heusler compounds FeRhCrSi and FePdCrSi: A first-Principles Study. *Appl. Sci.* **2018**, *8*, 2370. [CrossRef]

20. Schwarz, K. CrO$_2$ predicted as a half-metallic ferromagnet. *J. Phys. F Met. Phys.* **1986**, *16*, L211–L2015. [CrossRef]

21. Kim, W.; Kawaguchi, K.; Koshizaki, N. Fabrication and magnetoresistance of tunnel junctions using half-metallic Fe$_3$O$_4$. *J. Appl. Phys.* **2003**, *93*, 8032–8034. [CrossRef]

22. Huang, H.M.; Jiang, Z.Y.; Lin, Y.M.; Zhou, B.; Zhang, C.K. Design of half-metal and spin gapless semiconductor for spintronics application viacation substitution in methylammonium lead iodide. *Appl. Phys. Express* **2017**, *10*, 123002. [CrossRef]

23. Lv, S.H.; Li, H.P.; Liu, X.J.; Han, D.M.; Wu, Z.J.; Meng, J.A. new half-metallic ferromagnet La$_2$NiFeO$_6$: Predicted from first-principles calculations. *J. Phys. Chem. C* **2010**, *114*, 16710–16715. [CrossRef]

24. Haneef, M.; Arif, S.; Akbar, J.; Abdul-Malik, A. Theoretical investigations of half-metallicity in Cr-substituted GaN, GaP, GaAs, GaSb material systems. *J. Electron. Mater* **2014**, *43*, 3169–3176. [CrossRef]

25. Liu, H.; Zhang, J.M. Effect of two identical 3d transition-metal atoms M doping (M=V, Cr, Mn, Fe, Co, and Ni) on the structural, electronic, and magnetic properties of ZnO. *Phys. Status Solidi B* **2017**, *254*, 1700098. [CrossRef]

26. Gao, G.Y.; Yao, K.L.; Sasioglu, E.; Sandratskii, L.M.; Liu, Z.L.; Jiang, J.L. Half-metallic ferromagnetism in zinc-blende CaC, SrC, and BaC from first principles. *Phys. Rev. B* **2007**, *75*, 174442. [CrossRef]

27. Xie, W.H.; Xu, Y.Q.; Liu, B.G.; Pettifor, D.G. Half-metallic ferromagnetism and structural stability of zincblende phases of the transition-metal chalcogenides. *Phys. Rev. Lett.* **2003**, *91*, 037204. [CrossRef]

28. Yao, K.L.; Zhu, L.; Liu, Z.L. First-principles study of the ferromagnetic and half metallic properties of the fumarate-bridged polymer. *Eur. Phys. J. B* **2004**, *39*, 283–286. [CrossRef]

29. Huang, H.M.; Luo, G.Y.; Liu, G.Y.; Yao, K.L. First-principles study of electronic structure and half-metallicity of molecule-based ferromagnet Cr[N(CN)$_2$]$_2$. *Chin. J. Chem. Phys.* **2011**, *24*, 189–193. [CrossRef]

30. Zheng, J.M.; He, R.J.; Wan, Y.; Zhao, P.J.; Guo, P.; Jiang, Z.Y. Half- metal state of a Ti$_2$C monolayer by asymmetric surface decoration. *Phys. Chem. Chem. Phys.* **2019**, *21*, 3318–3326. [CrossRef] [PubMed]

31. Wang, Y.P.; Li, S.S.; Zhang, C.W.; Zhang, S.F.; Ji, W.X.; Li, P.; Wang, P.J. High-temperature Dirac half-metal PdCl$_3$: A promising candidate for realizing quantum anomalous Hall effect. *J. Mater. Chem. C* **2018**, *6*, 10284–10291. [CrossRef]

32. Lv, P.; Tang, G.; Yang, C.; Deng, J.M.; Liu, Y.Y.; Wang, X.Y.; Wang, X.Q.; Hong, J.W. Half-metallicity in two-dimensional Co$_2$Se$_3$ monolayer with superior mechanical flexibility. *2D Mater.* **2018**, *5*, 045026. [CrossRef]

33. Gao, Q.; Wang, H.L.; Zhang, L.F.; Hu, S.L.; Hu, Z.P. Computational study on the half-metallicity in transition metal-oxide-incorporated 2D g-C$_3$N$_4$ nanosheets. *Front. Phys.* **2018**, *13*, 138108. [CrossRef]

34. Deng, L.; Yang, Z.; Tan, L.; Zeng, L.; Zhu, Y.; Guo, L. Investigation of the Prussian Blue Analog Co$_3$[Co(CN)$_6$]$_2$ as an Anode Material for Nonaqueous Potassium-Ion Batteries. *Adv. Mater.* **2018**, *30*, 1802510. [CrossRef] [PubMed]

35. Sun, D.; Wang, H.; Deng, B.; Zhang, H.; Wang, L.; Wan, Q.; Yan, X.; Qu, M. A Mn Fe based Prussian blue Analogue@Reduced graphene oxide composite as high capacity and superior rate capability anode for lithium-ion batteries. *Carbon* **2019**, *143*, 706–713. [CrossRef]

36. Gao, Q.; Shi, N.; Sanson, A.; Sun, Y.; Milazzo, R.; Olivi, L.; Zhu, H.; Lapidus, S.H.; Zheng, L.; Chen, J.; Xing, X. Tunable Thermal Expansion from Negative, Zero, to Positive in Cubic Prussian Blue Analogues of GaFe(CN)$_6$. *Inorg. Chem.* **2018**, *57*, 14027–14030. [CrossRef] [PubMed]

37. Xiong, P.; Zeng, G.; Zeng, L.; Wei, M. Prussian blue analogues Mn[Fe(CN)$_6$]$_{0.6667}$·nH$_2$O cubes as an anode material for lithium-ion batteries. *Dalton Trans.* **2015**, *44*, 16746–16751. [CrossRef] [PubMed]

38. Holmes, S.H.; Girolami, G.S. Sol-Gel Synthesis of KV^{II}[Cr^{III}(CN)$_6$]·2H$_2$O: A Crystalline Molecule-Based Magnet with a Magnetic Ordering Temperature above 100 °C. *J. Am. Chem. Soc.* **1999**, *121*, 5593–5594. [CrossRef]

39. Sato, O.; Hayami, S.; Einaga, Y.; Gu, Z.Z. Control of the Magnetic and Optical Properties in Molecular Compounds by Electrochemical, Photochemical and Chemical Methods. *Bull. Chem. Soc. Jpn.* **2003**, *76*, 443–470. [CrossRef]

40. Wojdel, J.C.; Moreira, I.P.R.; Bromleyac, S.T.; Illas, F. Prediction of half-metallic conductivity in Prussian Blue derivatives. *J. Mater. Chem.* **2009**, *19*, 2032–2036. [CrossRef]

41. Blöchl, P.E. Projector augmented-wave method. *Phys. Rev. B* **1994**, *50*, 17953. [CrossRef]

42. Kresse, G.; Furthmüller, J. Efficiency of ab-initio total energy calculations for metals and semiconductors using a plane-wave basis set. *Comput. Mater. Sci.* **1996**, *6*, 15–50. [CrossRef]

43. Perdew, J.P.; Burke, K.; Ernzerhof, M. Generalized gradient approximation made simple. *Phys. Rev. Lett.* **1996**, *77*, 3865. [CrossRef] [PubMed]

44. Zhang, X.D.; Lv, Y.; Liu, C.; Wang, F.; Jiang, W. Site preference of transition-metal elements additions on mechanical and electronic properties of B$_2$DyCu-based alloys. *Mater. Des.* **2017**, *133*, 476–486. [CrossRef]

45. Zhang, X.D.; Huang, W.Y.; Ma, H.; Yu, H.; Jiang, W. First-principles prediction of the physical properties of ThM$_2$Al$_{20}$ (M= Ti, V, Cr) intermetallics. *Solid State Commun.* **2018**, *284–286*, 75–83. [CrossRef]

46. Mehmood, N.; Ahmad, R.; Murtaza, G. Ab initio investigations of structural, elastic, mechanical, electronic, magnetic, and optical properties of half-heusler compounds RhCrZ (Z = Si, Ge). *J. Supercond. Novel Magn.* **2017**, *30*, 2481–2488. [CrossRef]

47. Huang, Y.C.; Guo, X.F.; Ma, Y.L.; Shao, H.B.; Xiao, Z.B. Stabilities, electronic and elastic properties of L1$_2$-Al$_3$(Sc$_{1-x}$,Zr$_x$) with different Zr content: A first-principles study. *Physica B* **2018**, *548*, 27–33. [CrossRef]

48. Huang, H.M.; Luo, S.J.; Xiong, Y.C. Pressure-induced electronic, magnetic, half-metallic, and mechanical properties of half-Heusler compound CoCrBi. *J. Magn. Magn. Mater.* **2017**, *438*, 5–11. [CrossRef]

49. Anderson, O.L. A simplified method for calculating the debye temperature from elastic constants. *J. Phys. Chem. Solids* **1963**, *24*, 909–917. [CrossRef]

50. Allali, D.; Bouhemadou, A.; Zerarga, F.; Ghebouli, M.A.; Bin-Omran, S. Prediction study of the elastic and thermodynamic properties of the SnMg$_2$O$_4$, SnZn$_2$O$_4$ and SnCd$_2$O$_4$ spinel oxides. *Comput. Mater. Sci.* **2012**, *60*, 217–223. [CrossRef]

applied
sciences

MDPI

Article

Phase Stability and Magnetic Properties of Mn₃Z (Z = Al, Ga, In, Tl, Ge, Sn, Pb) Heusler Alloys

Haopeng Zhang [1], Wenbin Liu [1], Tingting Lin [1,2], Wenhong Wang [3] and Guodong Liu [1,*]

[1] School of Physics and Electronic Engineering, Chongqing Normal University, Chongqing 400044, China; hpzhang0201@126.com (H.Z.); wbliu1106@126.com (W.L.); tlin@tmm.tu-darmstadt.de (T.L.)
[2] Institute of Materials Science, Technische Universtät Darmstadt, 64287 Darmstadt, Germany
[3] Beijing National Laboratory for Condensed Matter Physics, Institute of Physics, Chinese Academy of Sciences, Beijing 100080, China; wenhong.wang@iphy.ac.cn
* Correspondence: gdliu1978@126.com

Received: 20 November 2018; Accepted: 5 December 2018; Published: 7 March 2019

Abstract: The structural stability and magnetic properties of the cubic and tetragonal phases of Mn₃Z (Z = Ga, In, Tl, Ge, Sn, Pb) Heusler alloys are studied by using first-principles calculations. It is found that with the increasing of the atomic radius of Z atom, the more stable phase varies from the cubic to the tetragonal structure. With increasing tetragonal distortion, the magnetic moments of Mn (A/C and B) atoms change in a regular way, which can be traced back to the change of the relative distance and the covalent hybridization between the atoms.

Keywords: phase stability; magnetic properties; covalent hybridization

1. Introduction

Tetragonal Heusler compounds have been receiving huge attention in recent years due to their potential applications in spintronic [1–5] and magnetoelectronic devices [6–9], such as ultrahigh density spintronic devices [9–13], spin-transfer torque (STT) [9–16] and permanent hard magnets [17,18]. Among the tetragonal Heusler compounds, Mn₃-based Heusler compounds exhibit very interesting properties. The previous theoretical and experimental studies [4,16,19–22] show that the tetragonal (DO₂₂) phase of Mn₃Ga compound is ferrimagnetic at room temperature and shows a unique combination of magnetic and electronic properties, including low magnetization, high uniaxial anisotropy, high spin polarization, and high Curie temperature. Because of these interesting properties, this material is believed to have potential for nanometer-sized spin transfer torque (STT) -based nonvolatile memories [4,16,23]. The first-principles calculations reveal that Mn₃Z (Z = Ga, Sn and Ge) type Heusler compounds can have three different structural phases, where each phase exhibits different magnetic properties [24]. There are also some other reports about the phase stability and the magnetic properties for these systems [25–28], but the relation between the phase stability and the magnetic properties of Mn₃Z tetragonal Heusler alloys has not been investigated in detail.

In this paper, the relation between the phase stability, magnetic properties, the covalent hybridization effect, and the relative position between atoms of Mn₃Z (Z = Al, Ga, In, Tl, Si, Ge, Sn, Pb) Heusler alloys has been investigated by using the first-principles calculations. It is found that the atomic radius of Z atoms and the level of distortion have great effects on the degree of the covalent hybridization between atoms in Mn₃Z system, which plays an important role in the phase stability and the magnetic properties of Mn₃Z Heusler compounds.

2. Calculation Details

The calculations of total energy, electronic structure, and magnetic moments were performed by the Cambridge Serial Total Energy Package (CASTEP) code based on the pseudopotential method with a plane-wave basis set [29]. The exchange and correlation effects were treated using the local density approximation (LDA) [30]. The plane wave basis set cut-off was 500 eV for all of the cases, and 182 k-points were employed in the irreducible Brillouin zone. The convergence tolerance for the calculations was selected as the difference in the total energy within the 1×10^{-6} eV/atom. These parameters ensure good convergences for the total energy.

3. Results and Discussion

Heusler alloys crystallize in a highly-ordered cubic structure, and have a stoichiometric composition of X_2YZ, where X and Y are transition-metal elements, and Z is a main group element. Generally, the Heusler structure can be considered as four interpenetrating f.c.c lattices along the space diagonal, in which the transition metal atoms occupy the A (0, 0, 0), B (0.25, 0.25, 0.25) and C (0.5, 0.5, 0.5) Wyckoff positions, respectively. The main group element occupies the D (0.75, 0.75, 0.75) position. The tetragonal Heusler alloys can be considered as tetragonal distortions of the cubic phase along the z direction, and the c/a ratio can be used to quantify the amount of tetragonal distortion [9,31,32]. For tetragonal Mn_3Z (Z = Al, Ga, In Tl, Si, Ge, Sn, Pb) alloys, the Mn(A), Mn(B), Mn(C) and Z atoms occupy the (0, 0, 0), (0.25, 0.25, 0.25), (0.5, 0.5, 0.5) and (0.75, 0.75, 0.75) Wyckoff positions, respectively.

As a typical example, we first present the results of Mn_3Ga alloy. Figure 1 shows the total energy as a function of c/a (ΔE_{total}-c/a curve) for Mn_3Ga alloy. The total energy of the cubic phase is set as the zero point. The lattice constant of the cubic phase is obtained by minimizing the total energy and is 5.66 Å. The unit cell volume is the same as that of cubic phase, and is fixed when the tetragonal distortion is considered. From Figure 1, it can be seen that there are two local energy minima on the ΔE_{total}-c/a curve, i.e., a shallow one is at c/a = 1.35 and a deeper one at c/a = 1. The latter is energetically favorable. Between the two energy minima, there is an energy barrier at c/a = 1.15. The lower and upper insets show the corresponding crystal structures, band structures, and densities of states (DOS) for the cubic (c/a = 1) and distorted (c/a = 1.35) cases. From the band structures, we can see that the cubic phase (c/a = 1) of Mn_3Ga is close to the half-metal, with a high degree of spin polarization. But in the tetragonal phase (c/a = 1.35), the band structure is completely different from the cubic one, and it can also be seen that the spin polarization declines rapidly at the Fermi level from the density of states patterns. This is mainly due to the fact that cubic symmetry is reduced after the tetragonal distortion.

Comparing the work reported by Delin Zhang et al. [31] with that by Claudia Felser et al. [33], it can be noted that the difference of volume can have a large impact on the ΔE_{total}-c/a curves. Therefore, we performed a series of investigations on the tetragonal distortion with different volumes to further understand the relation between the volume and the ΔE_{total}-c/a curves for Mn_3Ga alloy. In Figure 2a, we show the ΔE_{total}-c/a curves of Mn_3Ga alloy with different volumes (v = 17.0 nm^3, 18.0 nm^3, 19.0 nm^3, 20.0 nm^3, 21.0 nm^3, 22.0 nm^3), which correspond to the different lattice constants in the cubic phase. It can be seen that the shape of the ΔE_{total}-c/a curves varies with the change of the volume. There are two local energy minima in the ΔE_{total}-c/a curves for all the Mn_3Ga alloys with different volumes. For V = 17.0 nm^3 and 18.0 nm^3, the total energy of cubic phase is lower than that of the tetragonal phase, which indicates that the cubic phase is more stable than the tetragonal phase. As the volume expands to higher level than 19.0 nm^3, the total energy of the tetragonal phase becomes lower than the cubic phase. At same time, the energy barrier from the cubic phase to the tetragonal phase gradually decreases with the increasing volume, and finally, disappears at V = 21.0 nm^3. This indicates that the tetragonal phase becomes a more stable phase with the expanding volume, and it becomes easier to transform from the cubic phase to the tetragonal phase. So, from the ΔE_{total}-c/a curves with different volume of Mn_3Ga, it is clear that a very small change of the volume can lead to a great change of the shape of the ΔE_{total}-c/a curves. In other word, the phase stability of Mn_3Ga Heusler alloys is very sensitive to the change of the volume. For the Mn_3Ga systems, we can adjust the volume to achieve the alloys with different structures as well as possible martensitic transformations.

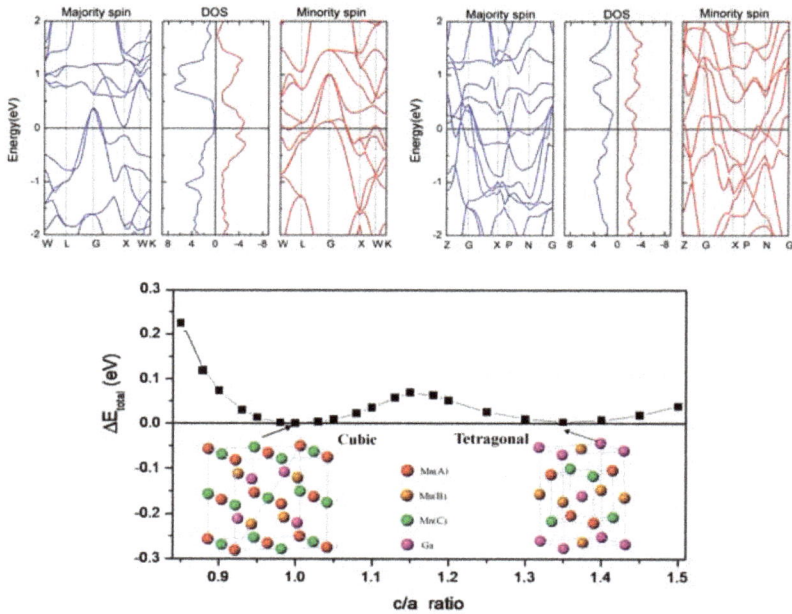

Figure 1. Total energy difference (per formula unit) relative to the cubic phase as a function of c/a for Mn$_3$Ga alloy ($\Delta E_{total} = E_{total}(c/a) - E_{total}(c/a = 1)$). The lower insets show the corresponding crystal structures and the upper insets show the band structures and densities of states for the cubic (c/a = 1) and distorted (c/a = 1.35) phases.

A good way to adjust the volume is to dope similar elements into the matrix. Therefore, next, we extend the research scope to all the other Mn$_3$Z (Z = Al, In, Tl, Si, Ge, Sn, Pb) alloys. We perform systematical investigations on ΔE_{total}-c/a curves for these alloys under their respective equilibrium cell volumes, which are achieved by their equilibrium lattice constant in the cubic structure. The equilibrium lattice constants in the cubic structure are 5.6 Å, 5.95 Å, 6.01 Å, 5.53 Å, 5.61 Å, 5.87 Å, and 6.01 Å for Mn$_3$Z (Z = Al, In, Tl, Si, Ge, Sn, Pb) alloys respectively. Their ΔE_{total}-c/a curves are shown in Figure 2b. For a clear analogy, the ΔE_{total}-c/a curve of Mn$_3$Ga is also replotted in Figure 2b, in which one can see that, similar to Mn$_3$Ga alloy, there are two local energy minima in the ΔE_{total}-c/a curves for all the other Mn$_3$Z alloys. One energy minimum is at c/a = 1 (cubic phase), and the other is at c/a = 1.35 (tetragonal phase), except for Mn$_3$Ge, where it is at c/a = 1.4. From Figure 2b, we can observe that when Z is cognate element, the ΔE_{total} of the tetragonal phase at c/a = 1.35 (for Mn$_3$Ge at c/a = 1.4) decreases gradually with the increase of atomic number. It is also clear that the smaller the atomic number, the smaller the volume of compound. The cubic phase is more stable than the tetragonal phase for the compounds with a small volume, such as Mn$_3$Al, Mn$_3$Ga, Mn$_3$Si, and Mn$_3$Ge. And the tetragonal phase is more stable in energy for the compounds with bigger atomic number, such as Mn$_3$In, Mn$_3$Tl, Mn$_3$Sn and Mn$_3$Pb.

The energy barrier between the two local energy minima is crucial to the occurrence of martensitic transformation (or reverse transformation) from the cubic (tetragonal) to tetragonal (cubic) phase. When the energy barrier is higher than the driving forces of phase transformation, the compound is stable in one of two local energy minima, and the martensitic transformation can not occur in the compound. Conversely, when the energy barrier is lower than the driving force of phase transformation, martensitic transformation may occur in the compound. In addition, from Figure 2b, it can be found that when the Z atom varies from Al to Tl (Si to Pb), the energy barrier exhibits a maximum at Mn$_3$Ga

(Mn$_3$Sn) for ΔE_{total}, and the local energy minimum of tetragonal phase changes from positive to negative value with the increasing atomic number.

All the above results imply that the atomic radius of the main group element Z has a great influence on the volume. We can mix different Z elements to obtain Mn$_3$Z alloys with the different volumes and stable phases. It should be noted that a thermoelastic martensitic transformation from cubic to tetragonal phase may also occur in the Mn$_3$Z alloys, since the energy barrier can be flexibly regulated by the mixture of different Z elements. So, the Mn$_3$Z alloys have the potential to be developed into a series of magnetic shape memory alloys originating from thermoelastic martensitic transformation.

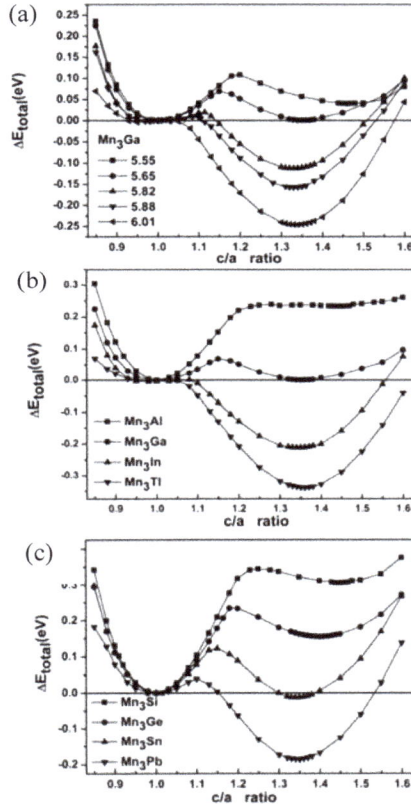

Figure 2. (**a**) Total energy as functions of c/a ratio for Mn$_3$Ga alloy with different volume (V = 17.0 nm^3, 18.0 nm^3, 19.0 nm^3, 20.0 nm^3, 21.0 nm^3, 22.0 nm^3). (**b**) Total energy difference (per formula unit) relative to the cubic phase as a function of c/a for Mn$_3$Z (Z = Al, Ga, In, Tl) and (**c**) for Mn$_3$Z (Z = Ge, Sn, Pb) alloys.

It is well known that with the increase of the distance between the main group Z atom and the nearest neighbor Mn(A) and Mn(C), the hybridization strength of the p-d orbitals between Z and Mn(A/C) atoms is weakened [34]. Before we start to analyze the magnetic properties, we perform an investigation on the change of the relative position of the atoms in Mn$_3$Z alloys during the tetragonal distortion. As shown in Figure 3a, the distance between Z and Mn(B) and the distance between Mn(A) and Mn(C) along the c axis increase linearly with the increase of the c/a ratio in the process of tetragonal distortion, while the distance between Z and Mn(B) and the distance between Mn(A) and Mn(C) along the a or b axis decreases linearly with the increase of the c/a ratio. Figure 3b show the changing curve of the distance between two nearest-neighbor atoms with the change of c/a ratio. It is clear that the distance of between two nearest neighbor atoms decreases first, and then increases with the increase of the c/a ratio and get a minimum at c/a = 1.

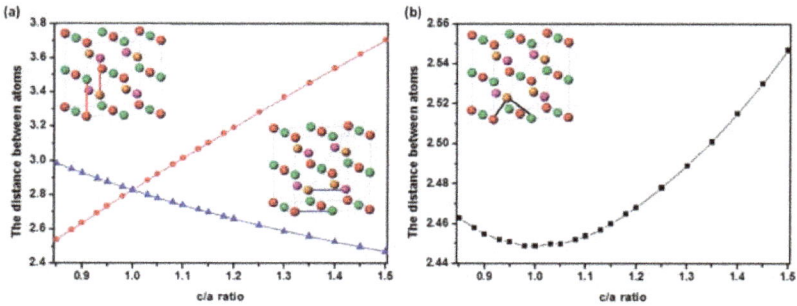

Figure 3. The change of interatomic distance in the process of tetragonal distortion for Mn_3Ga alloy (the lattice constant of cubic structure is 5.65 Å): (**a**) The red circle represents the distance between Ga and Mn(B) and the distance between Mn(A) and Mn(C) along c axis. The blue triangle represents the distance between Ga and Mn(B) and the distance between Mn(A) and Mn(C) along a or b axis. (**b**) The changing curve of the distance between two nearest neighbor atoms with the change of c/a. The thick lines between the atoms in the inset indicate the corresponding interatomic distance).

Next, we will compare and analyze the atomic magnetic moments for the Mn_3Ga alloy with different volumes, and all the other Mn_3Z alloys with the equilibrium volume, which are shown in Figures 4 and 5. As we know, with the increase of the lattice constant, the distance between atoms increases, which leads to the weakening of covalent hybridization between atoms. The weakened covalent hybridization will result in an increase of atomic magnetic moments in the Heusler alloys [35,36]. When we compress the lattice along the c-axis (c/a < 1), we see that the dependence of the magnetic moment of Mn(A/C) on c/a ratio shows three different tendency ranges with the change of the volume for Mn_3Ga alloy, as shown in Figure 4a. (1) When the volume is small (V = 17 nm^3), the magnetic moment of Mn(A/C) atom shows a sharp decrease with the increase of c/a. (2) When the volume is in the range of 18 nm^3~20 nm^3, the magnetic moment of Mn(A/C) almost remains constant with the increase of c/a, which indicates that the moment is very stable against the compressive strain along the c-axis. (3) When the volume is higher than 21 nm^3, the magnetic moment of Mn(A/C) increases slowly with the increase of c/a.

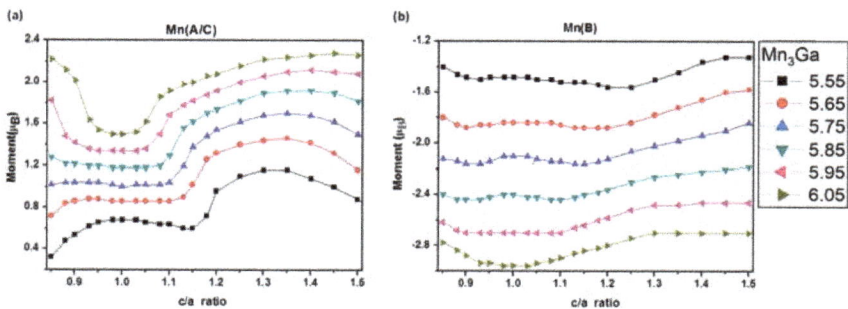

Figure 4. The atomic magnetic moments of Mn(A/C) (**a**) and Mn(B) (**b**) as functions of c/a ratio for Mn_3Ga alloy with different lattice constants.

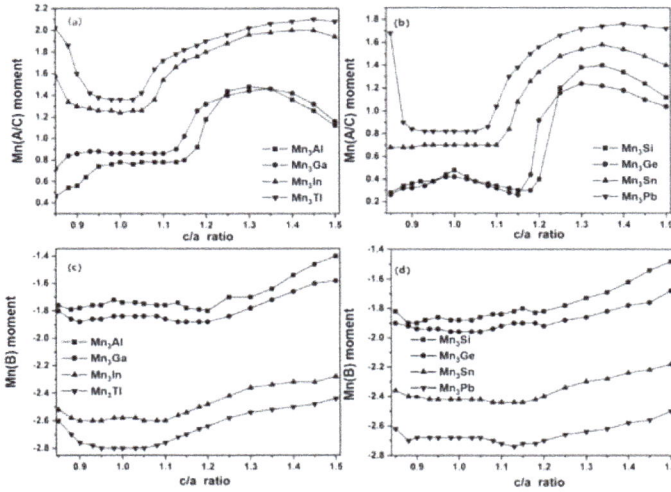

Figure 5. The atomic magnetic moments of Mn(A/C) (**a,b**) and Mn(B) (**c,d**) as functions of c/a ratio for Mn$_3$Z (Z = Al, Ga, In, Tl, Si, Ge, Sn, Pb) alloys.

The three different changing trends of the Mn(A/C) moment are due to the change of covalent hybridization between these atoms with the changing tetragonal distortion ratio [37]. There are three kinds of possible covalent hybridizations, i.e., between Ga and Mn(A/C), between Mn(A) and Mn(C) along c axis and between Mn(A/C) and Mn(B), which compete to determine the atomic magnetic moments. When c/a < 1, with the increase of the c/a ratio, the following occurs: (1) The distance between Ga and Mn(A/C) decreases, which leads to the strengthened p-d orbital covalent hybridization between Ga and Mn(A/C). Thus, the magnetic moment of Mn(A/C) decreases with the increase of c/a ratio. (2) The distance between Mn(A) and Mn(C) along the c axis increases, and the d-d orbital covalent hybridization between them was weakened. So, the magnetic moment of Mn(A/C) increases. (3) The distance between Mn(A/C) and Mn(B), namely, the distance between two nearest neighbor atoms, decreases and the d-d covalent hybridization is strengthened. Thus, the magnetic moments of Mn(A/C) and Mn(B) show a decrease trend. According to the above phenomena, it can be observed that when the volume is small (v = 17 nm^3), the d-d orbital covalent hybridization between Mn(A) and Mn(C) atoms plays a dominant role to lead to a sharp decrease of the magnetic moment of Mn(A/C) with the increase of c/a. When the volume is greater than 21 nm^3, the p-d orbital covalent hybridization between Ga and Mn(A/C) and the d-d orbital covalent hybridization between Mn(A/C) and Mn(B) atoms are the main contributors. As for the cases of 18~20 nm^3, which have moderate distances among atoms, the p-d orbital covalent hybridization between Ga and Mn(A/C), the d-d orbital covalent hybridization between Mn(A) and Mn(C) atoms, and the d-d orbital covalent hybridization between Mn(A/C) and Mn(B) atoms counteract each other. Thus, the magnetic moment of Mn(A/C) remains almost unchanged with the increase of c/a.

From Figure 4a, it can also be seen that when c/a > 1, the moment of Mn(A/C) first increases and then generates a downward trend. We might consider the case of c/a < 1 to understand the situation. Firstly, with the increase of c/a ratio, the distance between Mn(A/C) and Ga (also Mn(B)) increases. So, the p-d (d-d) covalent hybridization decreases and the moment of Mn(A/C) increases. Secondly, the distance between Mn(A) and Mn(C) atoms along c axis increases. Thus, the d-d covalent hybridization was weakened and the magnetic moments of Mn(A/C) increase. Thirdly, with the increase of c/a ratio, the distance between the Mn(A) and Mn(C) along the a or b axis decreases, which leads to the strengthened of d-d covalent hybridization between them and a decrease of the Mn(A/C) moment. So, we can know that the Mn(A/C) magnetic moment increases first, and then decreases with the increase of c/a, which may be attributed to the change of covalent hybridization originating from the change of interatomic distance. Furthermore, with the increase of volume, the position of the inflection point to go down gradually shifts to the right. This is because Mn_3Ga alloy with larger volume needs a larger degree of distortion (a larger c/a ratio) to make the distance between Mn(A) and Mn(C) along the a or b axes sufficiently small to achieve the same strength of d-d orbital covalent hybridization.

The magnetic moment of Mn(B) as a function of c/a ratio is plotted in Figure 4b for Mn_3Ga alloys with different volumes. In the process of tetragonal distortion, the distance between Ga and Mn(B) atoms along the c axis gradually increases, and the p-d orbital covalent hybridization between these atoms gradually weakens, which makes the Mn(B) moment continue to increase. At the same time, the distance between Ga and Mn(B) atom along the a or b axis gradually decreases with the increase of c/a ratio. So, the p-d orbital covalent hybridization between these atoms gradually strengthens and the Mn(B) moment decreases. Thus, we can also understand this changing behavior of the magnetic moment of Mn(B).

Furthermore, from Figure 4b, we can see that the interatomic distance effects on the covalent hybridization counteract each other when c/a ratio is small for Mn_3Ga alloys with V = 17.0~21.0 nm^3. And the magnetic moment of Mn(B) moment is essentially unchanged. But when c/a ratio increases to about 1.2, the p-d orbital covalent hybridization between Ga and Mn(B) atoms along the c axis plays a major role, and the magnetic moment of Mn(B) has an upward trend. When the volume increases to 22.0 nm^3, the distance between atoms is quite large, and the covalent hybridization becomes weaker. As c/a is in the range of 0.85–1, the p-d orbital covalent hybridization between Ga and Mn(B) atoms along the a or c axis and the d-d covalent hybridization between Mn(A/C) and Mn(B) are the main contributors, and the magnetic moment of Mn(B) has a downward trend. For the case of c/a > 1, both the p-d orbital covalent hybridization between Ga and Mn(B) along the c axis and the d-d covalent hybridization between Mn(A/C) and Mn(B) make the magnetic moment of Mn(B) show an upward trend.

We can see that the magnetic moments of Mn(A/C) and Mn(B) as functions of c/a ratio are very similar to Mn_3Ga alloy for all the other Mn_3Z (Z = Al, In, Tl, Si, Ge, Sn, Pb) alloys, as shown in Figure 5. As different main group elements have different atomic radii, the distance between atoms can be tuned by changing the main group element in Mn_3Z which is similar to that in Mn_3Ga alloy with different volume.

4. Conclusions

In summary, the structural and magnetic properties of tetragonal Heusler alloys Mn_3Z (Z = Ga, In, Tl, Ge, Sn, Pb) have been systemically investigated by the first-principles calculations. The calculations indicate that the stability of the system is very sensitive to changes of volume. And the volume can be tuned by changing the main group element in Mn_3Z alloys. The p-d and d-d covalent hybridization play very important roles during the tetragonal distortion, and have great influence on the atomic magnetic moments in Mn_3Z alloys.

Author Contributions: Conceptualization, G.L. and H.Z.; methodology, H.Z.; formal analysis, G.L., H.Z. and W.W.; investigation, H.Z., W.L. and T.L.; writing—original draft preparation, H.Z.

Funding: This work was supported by the Program for Leading Talents in Science and Technology Innovation of Chongqing City (No. CSTCKJCXLJRC19) and Natural Science Foundation of Tianjin City (No. 16JCYBJC17200).

Conflicts of Interest: The authors declare no conflict of interest.

References

1. Wollmann, L.; Fecher, G.H.; Chadov, S.; Felser, C. A scheme for spin-selective electron localization in Mn_3Ga Heusler material. *J. Phys. D Appl. Phys.* **2015**, *48*, 164004. [CrossRef]
2. Wu, H.K.F.; Sudoh, I.; Xu, R.H.; Si, W.S.; Vaz, C.A.F.; Kim, J.; Vallejo-Fernandez, G.; Hirohata, A. Large exchange bias induced by polycrystalline Mn_3Ga antiferromagnetic films with controlled layer thickness. *J. Phys. D Appl. Phys.* **2018**, *51*, 215003. [CrossRef]
3. Wu, F.; Mizukami, S.; Watanabe, D.; Sajitha, E.P.; Naganuma, H.; Oogane, M.; Ando, Y.; Miyazaki, T. Structural and Magnetic Properties of Perpendicular Magnetized MnGa Epitaxial Films. *IEEE Trans. Magn.* **2010**, *46*, 1863–1865. [CrossRef]
4. Winterlik, J.; Balke, B.; Fecher, G.H.; Felser, C. Structural, electronic, and magnetic properties of tetragonal $Mn_{3-x}Ga$: Experiments and first-principles calculations. *Phys. Rev. B* **2008**, *77*, 054406. [CrossRef]
5. Feng, W.W.; Thiet, D.V.; Dung, D.D.; Shin, Y.; Cho, S. Substrate-modified ferrimagnetism in MnGa films. *J. Appl. Phys.* **2010**, *108*, 113903. [CrossRef]
6. Niida, H.; Hori, T.; Onodera, H.; Yamaguchi, Y.; Nakagawa, Y. Magnetization and coercivity of $Mn_{3-\delta}Ga$ alloys with a $D0_{22}$-type structure. *J. Appl. Phys.* **1996**, *79*, 5946–5948. [CrossRef]
7. Elm, M.T.; Hara, S. Transport Properties of Hybrids with Ferromagnetic MnAs Nanoclusters and Their Potential for New Magnetoelectronic Devices. *Adv. Mater.* **2014**, *26*, 8079–8095. [CrossRef]
8. Ma, Q.L.; Kubota, T.; Mizukami, S.; Zhang, X.M.; Naganuma, H.; Oogane, M.; Ando, Y.; Miyazaki, T. Interface tailoring effect on magnetic properties and their utilization in MnGa-based perpendicular magnetic tunnel junctions. *Phys. Rev. B* **2013**, *87*, 184426. [CrossRef]
9. Winterlik, J.; Chadov, S.; Gupta, A.; Alijani, V.; Gasi, T.; Filsinger, K.; Balke, B.; Fecher, G.H.; Jenkins, C.A.; Casper, F.; et al. Design Scheme of New Tetragonal Heusler Compounds for Spin-Transfer Torque Applications and its Experimental Realization. *Adv. Mater.* **2012**, *24*, 6283–6287. [CrossRef]
10. Gutierrez-Perez, R.M.; Holguın-Momaca, J.T.; Elizalde-Galindo, J.T.; Espinosa-Maga~na, F.; Olive-Mendez, S.F. Giant magnetization on Mn_3Ga ultra-thin films grown by magnetron sputtering on SiO_2/Si(001). *J. Appl. Phys.* **2015**, *117*, 123902. [CrossRef]
11. Baltz, V.; Manchon, A.; Tsoi, M.; Moriyama, T.; Ono, T.; Tserkovnyak, Y. Antiferromagnetic spintronics. *Rev. Mod. Phys.* **2018**, *90*, 015005. [CrossRef]
12. Kurt, H.; Baadji, N.; Rode, K.; Venkatesan, M.; Stamenov, P.; Sanvito, S.; Coey, J.M.D. Magnetic and electronic properties of $D0_{22}$-Mn_3Ge (001) films. *Appl. Phys. Lett.* **2012**, *101*, 132410. [CrossRef]
13. Wu, F.; Sajitha, E.P.; Mizukami, S.; Watanabe, D.; Miyazaki, T.; Naganuma, H.; Oogane, M.; Ando, Y. Electrical transport properties of perpendicular magnetized Mn-Ga epitaxial films. *Appl. Phys. Lett.* **2010**, *96*, 042505. [CrossRef]
14. Sugihara, A.; Suzuki, K.Z.; Miyazaki, T.; Mizukami, S. Magnetic properties of ultrathin tetragonal Heusler $D0_{22}$-Mn_3Ge perpendicular-magnetized films. *J. Appl. Phys.* **2015**, *117*, 17B511. [CrossRef]
15. Bai, Z.Q.; Cai, Y.Q.; Shen, L.; Yang, M.; Ko, V.; Han, G.C.; Feng, Y.P. Magnetic and transport properties of $Mn_{3-x}Ga$/MgO/$Mn_{3-x}Ga$ magnetic tunnel junctions: A first-principles study. *Appl. Phys. Lett.* **2012**, *100*, 022408. [CrossRef]
16. Balke, B.; Fecher, G.H.; Winterlik, J.; Felser, C. Mn_3Ga, a compensated ferrimagnet with high Curie temperature and low magnetic moment for spin torque transfer applications. *Appl. Phys. Lett.* **2007**, *90*, 152504. [CrossRef]
17. Zhu, L.J.; Nie, S.H.; Meng, K.K.; Pan, D.; Zhao, J.H.; Zheng, H.Z. Multifunctional $L1_0$-$Mn_{1.5}Ga$ Films with Ultrahigh Coercivity, Giant Perpendicular Magnetocrystalline Anisotropy and Large Magnetic Energy Product. *Adv. Mater.* **2012**, *24*, 4547–4551. [CrossRef]
18. Zhu, L.J.; Pan, D.; Nie, S.H.; Lu, J.; Zhao, J.H. Tailoring magnetism of multifunctional Mn_xGa films with giant perpendicular anisotropy. *Appl. Phys. Lett.* **2013**, *102*, 132403. [CrossRef]
19. Kurt, H.; Rode, K.; Venkatesan, M.; Stamenov, P.; Coey, J.M.D. High spin polarization in epitaxial films of ferrimagnetic Mn_3Ga. *Phys. Rev. B* **2011**, *83*, 020405. [CrossRef]
20. Chadov, S.; Souza, S.W.D.; Wollmann, L.; Kiss, J.; Fecher, G.H.; Felser, C. Chemical disorder as an engineering tool for spin polarization in Mn_3Ga-based Heusler systems. *Phys. Rev. B* **2015**, *91*, 094203. [CrossRef]
21. Khmelevskyi, S.; Shick, A.B.; Mohn, P. Prospect for tunneling anisotropic magneto-resistance in ferrimagnets: Spin-orbit coupling effects in Mn_3Ge and Mn_3Ga. *Appl. Phys. Lett.* **2016**, *109*, 222402. [CrossRef]

22. You, Y.R.; Xu, G.Z.; Hu, F.; Gong, Y.Y.; Liu, E.; Peng, G.; Xu, F. Designing magnetic compensated states in tetragonal Mn₃Ge-based Heusler alloys. *J. Magn. Magn. Mater.* **2017**, *429*, 40–44. [CrossRef]
23. Jeong, J.; Ferrante, Y.; Faleev, S.V.; Samant, M.G.; Felser, C.; Parkin, S.S.P. Termination layer compensated tunnelling magnetoresistance in ferrimagnetic Heusler compounds with high perpendicular magnetic anisotropy. *Nat. Commun.* **2016**, *7*, 10276. [CrossRef] [PubMed]
24. Zhang, D.L.; Yan, B.H.; Wu, S.C.; Kübler, J.; Kreiner, G.; Parkin, S.S.P.; Felser, C. First-principles study of the structural stability of cubic, tetragonal and hexagonal phases in Mn₃Z (Z = Ga, Sn and Ge) Heusler compounds. *J. Phys. Condens. Matter.* **2013**, *25*, 206006. [CrossRef] [PubMed]
25. Yun, W.S.; Cha, G.B.; Kim, I.G.; Rhim, S.H.; Hong, S.C. Strong perpendicular magnetocrystalline anisotropy of bulk and the (001) surface of DO₂₂ Mn₃Ga: A density functional study. *J. Phys. Condens. Matter* **2012**, *24*, 416003. [CrossRef] [PubMed]
26. Minakuchi, K.; Umetsu, R.Y.; Ishida, K.; Kainuma, R. Phase equilibria in the Mn-rich portion of Mn–Ga binary system. *J. Alloys Compd.* **2012**, *537*, 332–337. [CrossRef]
27. Cable, J.W.; Wakabayashi, N.; Radhakrishna, P. Magnetic excitations in the triangular antiferromagnets Mn₃Sn and Mn₃Ge. *Phys. Rev. B* **1993**, *48*, 6159–6166. [CrossRef]
28. Sung, N.H.; Ronning, F.; Thompson, J.D.; Bauer, E.D. Magnetic phase dependence of the anomalous Hall effect in Mn₃Sn single crystals. *Appl. Phys. Lett.* **2018**, *112*, 132406. [CrossRef]
29. Segall, M.D.; Lindan, P.J.D.; Probert, M.J.; Pickard, C.J.; Hasnip, P.J.; Clark, S.J.; Payne, M.C. First-principles simulation: Ideas, illustrations and the CASTEP code. *J. Phys. Condens. Matter.* **2002**, *14*, 2717–2744. [CrossRef]
30. Vosko, S.H.; Wilk, L.; Nusair, M. Accurate spin-dependent electron liquid correlation energies for local spin density calculations: A critical analysis. *Can. J. Phys.* **1980**, *58*, 1200–1211. [CrossRef]
31. Ye, M.; Kimura, A.; Miura, Y.; Shirai, M.; Cui, Y.T.; Shimada, K.; Namatame, H.; Taniguchi, M.; Ueda, S.; Kobayashi, K.; et al. Role of Electronic Structure in the Martensitic Phase Transition of Ni₂Mn₁₊ₓSn₁₋ₓ Studied by Hard-X-Ray Photoelectron Spectroscopy and Ab Initio Calculation. *Phys. Rev. Lett.* **2010**, *104*, 176401. [CrossRef] [PubMed]
32. Siewert, M.; Gruner, M.E.; Dannenberg, A.; Chakrabarti, A.; Herper, H.C.; Wuttig, M.; Barman, S.R.; Singh, S.; Al-Zubi, A.; Hickel, T.; et al. Designing shape-memory Heusler alloys from first-principles. *Appl. Phys. Lett.* **2011**, *99*, 191904. [CrossRef]
33. Felser, C.; Alijani, V.; Winterlik, J.; Chadov, S.; Nayak, A.K. Tetragonal Heusler Compounds for Spintronics. *IEEE Trans. Magn.* **2013**, *49*, 682–685. [CrossRef]
34. Liu, G.D.; Dai, X.F.; Liu, H.Y.; Chen, J.L.; Li, Y.X.; Xiao, G.; Wu, G.H. Mn₂CoZ (Z = Al, Ga, In, Si, Ge, Sn, Sb) compounds: Structural, electronic, and magnetic properties. *Phys. Rev. B* **2008**, *77*, 014424. [CrossRef]
35. Li, G.J.; Liu, E.K.; Zhang, H.G.; Zhang, Y.J.; Xu, G.Z.; Luo, H.Z.; Zhang, H.W.; Wang, W.H.; Wu, G.H. Role of covalent hybridization in the martensitic structure and magnetic properties of shape-memory alloys: The case of Ni₅₀Mn₅₊ₓGa₃₅₋ₓCu₁₀. *Appl. Phys. Lett.* **2013**, *102*, 062407. [CrossRef]
36. Zhang, Y.J.; Wang, W.H.; Zhang, H.G.; Liu, E.K.; Ma, R.S.; Wu, G.H. Structure and magnetic properties of Fe₂NiZ (Z = Al, Ga, Si and Ge) Heusler alloys. *Physica B* **2013**, *420*, 86–89. [CrossRef]
37. Liu, E.K.; Wang, W.H.; Feng, L.; Zhu, W.; Li, G.J.; Chen, J.L.; Zhang, H.W.; Wu, G.H.; Jiang, C.B.; Xu, H.B.; et al. Stable magnetostructural coupling with tunable magnetoresponsive effects in hexagonal ferromagnets. *Nat. Commun.* **2012**, *3*, 873. [CrossRef] [PubMed]

applied sciences

MDPI

Article

First-Principles Prediction of Skyrmionic Phase Behavior in GdFe$_2$ Films Capped by $4d$ and $5d$ Transition Metals

Soyoung Jekal [1,2,*]**, Andreas Danilo** [3]**, Dao Phuong** [4] **and Xiao Zheng** [4]

1   Laboratory of Metal Physics and Technology, Department of Materials, ETH Zurich, 8093 Zurich, Switzerland
2   Condensed Matter Theory Group, Paul Scherrer Institute, CH-5232 Villigen PSI, Switzerland
3   Laboratory for Solid State Physics, Department of Physics, ETH Zurich, 8093 Zurich, Switzerland; a.danilo@gmail.com
4   Hefei National Laboratory, University of Science and Technology of China, Hefei 230026, Anhui, China; d.phuong@gmail.com (D.P.); xiaozz@gmail.com (X.Z.)
*   Correspondence: so-young.jekal@mat.ethz.ch; Tel.: +41-44-632-26-43

Received: 22 January 2019; Accepted: 8 February 2019; Published: 13 February 2019

Abstract: In atomic GdFe$_2$ films capped by $4d$ and $5d$ transition metals, we show that skyrmions with diameters smaller than 12 nm can emerge. The Dzyaloshinskii–Moriya interaction (DMI), exchange energy, and the magnetocrystalline anisotropy (MCA) energy were investigated based on density functional theory. Since DMI and MCA are caused by spin–orbit coupling (SOC), they are increased with $5d$ capping layers which exhibit strong SOC strength. We discover a skyrmion phase by using atomistic spin dynamic simulations at small magnetic fields of ∼1 T. In addition, a ground state that a spin spiral phase is remained even at zero magnetic field for both films with $4d$ and $5d$ capping layers.

Keywords: skyrmion; Dzyaloshinskii–Moriya interaction; exchange energy; magnetic anisotropy

---

1. Introduction

In the sphere of magnetic memory storage (especially in spintronics), magnetic skyrmions, which are localized topologically protected spin structures, are promising candidates due to their unique properties [1–3]. Even though skyrmions have long been investigated by simulations such as micromagnetic and phenomenological model calculations [4–6], the experimental discovery of skyrmions was came about very recently in bulk MnSi [7]. Since then, researchers have focused on observing stabilized skyrmions experimentally in not only bulk crystals [8,9], but also thin films and multilayers [10–14].

At room temperature, Neél-type skyrmions with a diameter of ∼50 nm are found in multilayer stacks, such as Pt/Co/Ta and Ir/Fe/Co/Pt [15,16]. However, to use them in memory and logic devices, a further reduction in skyrmion sizes is necessary. As a result of the decreasing stability of small skyrmions at room temperature, thicker magnetic layers are required to increase stability [17,18]. For multilayer systems consisting of ferromagnet and heavy metals, interfacial anisotropy and the strength of Dzyaloshinskii–Moriya interaction (DMI) reduces as the thickness of ferromagnetic layer increases. Moreover, the skyrmion Hall effect is a challenge when it comes to moving skyrmions in electronics devices [19–21]. Amorphous rare-earth–transition-metal (RE–TM) ferrimagnets are one of the potential materials to overcome these challenges. Their Intrinsic perpendicular magnetocrystalline anisotropy (MCA) gives an advantage in stabilizing skyrmions by using relatively thick magnetic layers (∼5 nm) [22]. Another advantage of RE–TM alloys is that the skyrmion Hall effect is largely reduced by the near zero magnetization of RE–TM alloys [23]. Furthermore, in perspective of the applications, all-optical helicity-dependent switching (AO-HDS) has been shown in RE–TM alloys due to its ultrafast switching. Recently, AO-HDS has been demonstrated in RE–TM alloys using a circularly polarized laser. As a result, RE–TM alloys have drawn interest in the field of skyrmions research.

In recent, large skyrmions with diameter of ∼150 nm have been observed in Pt/GdFeCo/MgO [24], and skyrmion bound pairs are found in Gd/Fe multilayers [25]. However, further tuning is essential to reduce the size of skyrmions in RE–TM alloys.

In the present paper, magnetic properties such as DMI, MCA, and magnetic phase transition are investigated in atomic GdFe$_2$ films capped by 4d and 5d transition metals (TMs) using first principles density functional theory (DFT) calculations and atomistic spin dynamics simulations. We recognize that the 5d TMs give rise to a large DMI and strong MCA due to their large spin–orbit coupling (SOC) and orbital hybridization with 3d bands of Fe atom. Firstly, an extended Heisenberg model is studied by using atomistic spin dynamics. Then, we parameterize an extended Heisenberg model from DFT calculations. According to the phase diagram observed at zero temperature, there are phase transitions under externally applied magnetic fields of the order of ∼1 T. The magnetic phase changes from the spin spiral state to the ferromagnetic state via skyrmion lattice, the diameters of isolated skyrmions amount to 6 to 15 nm depending on the capping layers.

2. Methods

We used DFT as implemented in the Quantum Espresso [26] and Fleur code [27] to investigate the electronic and magnetic properties of GdFe$_2$/TMs film. For the TMs capping layers, we have considered Ru, Rh, Pd, and Ag in 4d and Os, Ir, Pt and Au in 5d. For the exchange–correlation potential we adapted the generalized gradient approximation (GGA). The wave functions were expanded by a plane-wave basis set with an optimized cutoff energy of 350 Ry, and the Brillouin zone was sampled via a 12 × 12 × 1 k-point mesh. Different mesh values from 36 to 256 were tested to ensure the precision of our calculations, with the convergence criterion being 0.1 μeV. The convergence with respect to cutoff was also carefully checked.

Total energy E(q) is calculated along the paths of Γ̄-K̄ and Γ̄-M̄ which have the highest symmetry among other directions in the two-dimensional Brillouin zone (2D BZ). E(q) with and without SOC [28] are separately displayed in Figure 1. In the 2D BZ, we characterize spin spiral phase using the wave vector **q** with a constant angle of ϕ, where ϕ is defined as **q**·**R**.

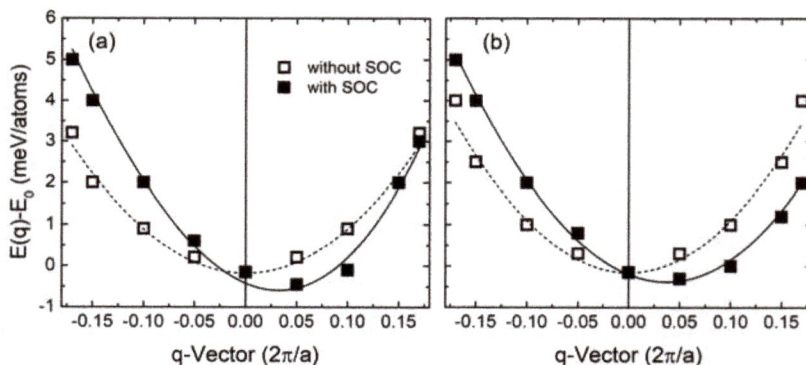

Figure 1. Energy dispersion E(q) of homogeneous cycloidal flat spin spirals in high-symmetry direction Γ̄-K̄ for (**a**) GdFe$_2$/Rh and (**b**) GdFe$_2$/Rh films. Filled and empty symbols represent E(q) with and without SOC, respectively. The energy is given relative to the magnetic ground state. The dispersion is fitted to the Heisenberg model (dotted line) and includes the DMI and MCA (solid line).

In order to examine the magnetically characteristic of $GdFe_2$ films with TM capping layers, we adopt the atomistic spin model given by References [29–31]:

$$H = -\sum_{ij} J_{ij}(m_i \cdot m_j) - \sum_{ij} D_{ij}(m_i \times m_j) + \sum_i K(m_i^z)^2 - \sum_i \mu_s(B \cdot m_i). \tag{1}$$

By using Equation (1), we can describe the magnetic interactions between two neighbor Fe atoms with spins of $\mathbf{M}_i$ and $\mathbf{M}_j$ at sites $\mathbf{R}_i$ and $\mathbf{R}_j$, respectively. Here, m_i is defined as $\mathbf{M}_i/\mu_s$. Both energy dispersion curves (with and without SOC) are calculated and fitted to extract the parameters for the exchange interactions (J_{ij}) and the DMI (D_{ij}).

We then compute the magnetic state by solving the Landau–Lifshitz–Gilbert (LLG) equation,

$$\frac{d\mathbf{S}_i}{dt} = -\gamma' \mathbf{S}_i \times (\mathbf{B}_i^{\text{eff}} + \mathbf{B}_i^{\text{th}}) - \gamma' \alpha \mathbf{S}_i \times [\mathbf{S}_i \times (\mathbf{B}_i^{\text{eff}} + \mathbf{B}_i^{\text{th}})]. \tag{2}$$

Here α denotes the Gilbert damping parameter. When γ is the gyromagnetic ratio, γ' represents $\frac{\gamma}{1+\alpha^2}$. $\mathbf{B}_i^{\text{eff}}$ is the effective magnetic field at site i, and $\mathbf{B}_i^{\text{th}}$ is the thermal noise. The LLG simulations were done with mumax3 [32]. For the present systems we use material parameters obtained from DFT: K = 2–14 meV and D = 0.2–1.6 meV (see Figure 2). To verify the numerical stability of the simulations, calculations with different cell sizes were performed. Finally, the thin films are discretized in a $400 \times 400 \times 2$ mesh with periodic boundary conditions in in-plane directions.

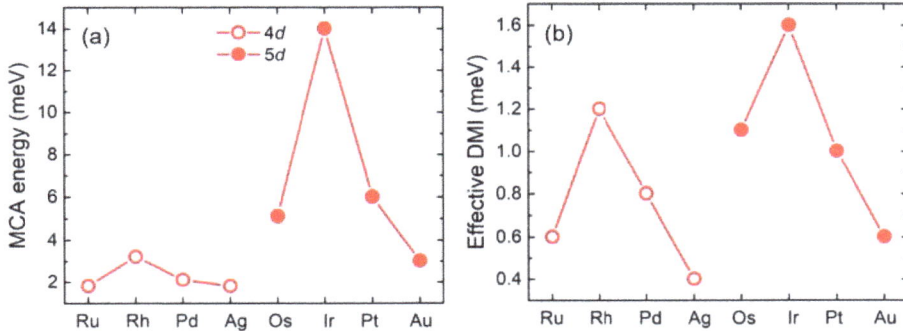

Figure 2. (a) Total magnetocrystalline anisotropy (MCA) energy and (b) effective Dzyaloshinskii–Moriya interaction (DMI) of $GdFe_2$ with TM capping layer.

The MCA energy was calculated using the force theorem and defined as the total energy difference between the magnetization perpendicular to the [100]-plane and parallel to the [100]-plane. Therefore, MCA energy $E_{\text{MCA}} = E_{[100]} - E_{[001]}$, where $E_{[100]}$ and $E_{[001]}$ are the total energies with the magnetization aligned along the [100] and [001] of the magnetic anisotropy, respectively.

3. Results and Discussion

The in-plane lattice constant of 7.32 Å was taken from the experimental lattice constant of Laves phase of $GdFe_2$, with lattice mismatches of 3.6% (Rh)–14.2% (Os), as depicted in Figure 3a. From the total energy calculation, it was confirmed that the hollow site is the most energetically favorable to stack the TM layer (see Figure 3). The atoms of $GdFe_2$ and TM capping layer were fully relaxed by atomic force calculations.

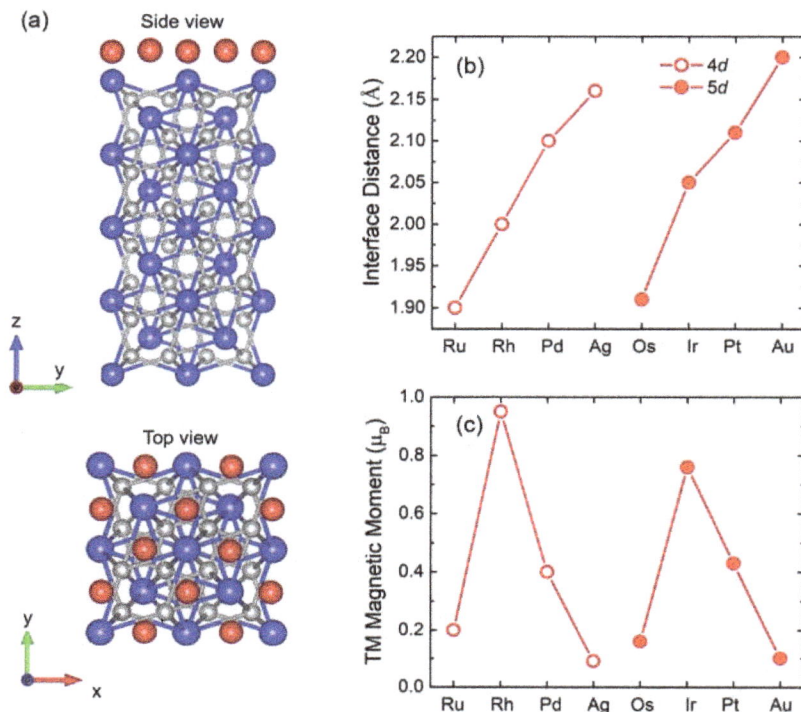

Figure 3. (**a**) Side view and top view of GeFe$_2$ film capped by a transition-metal (TM) monolayer. Blue, gray, and red balls represent Gd, Fe, and TM atoms, respectively. TM atoms are on the hollow site of GeFe$_2$; (**b**) Interface distances between the TM capping layer and GeFe$_2$ after structural optimization; (**c**) Magnetic moments of TM atoms, induced by GeFe$_2$.

After structural optimization, the interface distances between the TM capping layer and the GdFe$_2$ is presented in Figure 3b. As the atomic number becomes larger in the 4d and 5d TMs , the interlayer distances increase monotonically. Induced spin moments of the TMs for TM/GdFe$_2$ are presented in Figure 3c. The Rh and Ir capping layers, which are the Co-group elements, are found to have the largest moments of 0.98 and 0.80 μ_B. For all of the TM/GdFe$_2$, the direction of magnetization is favored to perpendicularly orientate to the film plane. Interestingly, the MCA energy and DMI of GdFe$_2$ films capped by 5d TMs are significantly larger than those of GdFe$_2$ with 4d TMs. In particular, the Ir-capped GdFe$_2$ film exhibits the largest MCA energy of 14.1 meV and effective DMI of 1.6 meV. We attribute the substantial enhancement of MCA energy and DMI in GdFe$_2$ with the 5d capping layer to the strong SOC of the 5d orbitals because the SOC is proportional to the fourth power of the atomic number. Since the 4d also exhibit similar trend with 5d, Rh has the largest magnetic moments and MCA energy among other 4d TMs. This is related to the band-filling effect and orbital hybridization.

The calculated energy dispersion E(q) of spin spirals is presented in Figure 1 along the high-symmetry direction, $\bar{\Gamma}$-$\bar{K}$ for GdFe$_2$ capped by Rh and Ir which exhibit the largest magnetic moment, MCA energy, and effective DMI among the 4d and 5d TM elements, respectively. In the results without SOC, a minimum point of the energy dispersion is observed at the $\bar{\Gamma}$ point, and it degenerates for right-($q > 0$) and left-rotating ($q < 0$) spirals. For both Rh- and Ir-capped films, it is confirmed that the out-of-plane direction is an easy magnetization axis due to SOC (see Figure 2a). As a result of imperfect inversion symmetry at the interface, the SOC for spin spirals derives DMI in system [33,34]. Therefore, DMI leads to non-collinear spin structures with the magnetic moments on an oblique angle. In case of the inclusion of the DMI, the E(q) has the lowest value for a homogeneous cycloidal flat spin spiral state with a particular rotational sense [35]. As presented in Figure 1, an energy minimum of −0.50 meV/atom and −0.35 meV/atom compared to the ground magnetic state appears for a right-rotating spin spiral for $GdFe_2$ films with Rh and Ir capping, respectively.

A skyrmion can be considered to be an intermediate state between spin spiral state and ferromagnetic state in a magnetic material because it rises from the competition between the exchange interaction that is responsible for the ferromagnetic state and the anisotropic exchange that generates spin spiral behavior. To investigate the magnetic phase transitions in $GdFe_2$/Rh and $GdFe_2$/Ir under the external magnetic field at 0 K, we have performed atomistic spin-dynamics simulations using the model described by Equation (1). Using the parameters obtained from DFT, the magnetic phase diagrams is displayed in Figure 4a,b. For both films capped by Rh and Ir, the ground magnetic state is a spin spiral consistent with the energy minimum at zero applied magnetic field. However, for the film capped by Rh, the skyrmion lattice is energetically stable at a critical field value of ~1.12 T, and this skyrmion lattice phase is changed to the ferromagnetic phase by a larger critical field value of ~2.25 T. For the film capped by Ir, the skyrmion lattice emerges at relatively weak field of 0.75 T, and disappears for a large filed of ~1.74 T.

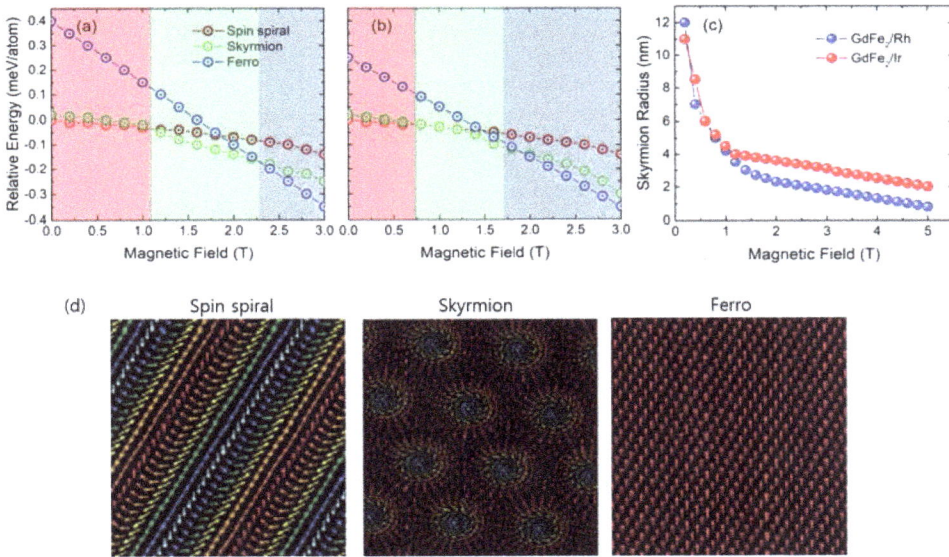

Figure 4. Phase diagrams for the (**a**) $GdFe_2$/Rh and the (**b**) $GdFe_2$/Ir films at zero temperature. The relative energies of the spin spiral states, skyrmion lattice, and ferromagnetic state are shown. The red, green, and blue colors represent the regime of the spin spiral states, skyrmion lattice, and ferromagnetic state, respectively. (**c**) Radii of skyrmions in the films of $GdFe_2$/Rh and $GdFe_2$/Ir as a function of the applied magnetic field. (**d**) Schematic representation of possible spin configurations in a magnetic material with Dzyaloshinsky–Moriya interaction for different values of an external field.

In our simulation, the spin structure is relaxed using spin dynamics. As shown in Figure 4c, skyrmions with a diameter of ~2–4 nm emerge under external magnetic fields of 1–2 T for both Rh- and Ir-capped $GdFe_2$. The size of skyrmions decreases rapidly with the increasing value of applied magnetic field. For deeper insights into the skyrmion size, the diameter has been computed for isolated single skyrmions in two different ways: (i) Using the fixed MCA energy and exchange constants obtained from DFT calculation but varying the DMI value; (ii) using fixed DMI obtained from DFT but varying the MCA. From these calculations we confirmed that the skyrmion size decreases with reduced DMI but it expands with reduced MCA.

4. Conclusions

The creation of extremely small, isolated and stabilized skyrmions of sizes of few nanometers in GdFe$_2$ films can be predicted by 4d and 5d TMs capping. While the atomistic spin model behavior was studied by spin dynamics simulations, first-principles parameters were obtained from density functional theory calculations. For future experimental work, this simulation work guides us in the exploration of novel skyrmion systems.

Author Contributions: conceptualization, S.J.; methodology, S.J. and A.D.; data curation, D.P. and X.Z.; writing draft, S.J., A.D., D.P., and X.Z.; project administration, S.J. and A.D.

Funding: This research was funded by ETH Zürich central funding.

Conflicts of Interest: The authors declare no conflict of interest.

References

1. Fert, A.; Cros, V.; Sampaio, J. Skyrmions on the track. *Nat. Nanotechnol.* **2013**, *8*, 152. [CrossRef]
2. Wiesendanger, R. Nanoscale magnetic skyrmions in metallic films and multilayers: A new twist for spintronics. *Nat. Rev. Mater.* **2016**, *1*, 16044. [CrossRef]
3. Kiselev, N.; Bogdanov, A.; Schäfer, R.; Rößler, U. Chiral skyrmions in thin magnetic films: New objects for magnetic storage technologies? *J. Phys. D Appl. Phys.* **2011**, *44*, 392001. [CrossRef]
4. Bogdanov, A.N.; Yablonskii, D.A. Thermodynamically stable "vortices" in magnetically ordered crystals. The mixed state of magnets *Sov. Phys. JETP* **1989**, *68*, 101.
5. Bogdanov, A.; Hubert, A. Thermodynamically stable magnetic vortex states in magnetic crystals. *J. Magn. Magn. Mater.* **1994**, *138*, 255–269. [CrossRef]
6. Bogdanov, A.; Rößler, U. Chiral symmetry breaking in magnetic thin films and multilayers. *Phys. Rev. Lett.* **2001**, *87*, 037203. [CrossRef]
7. Mühlbauer, S.; Binz, B.; Jonietz, F.; Pfleiderer, C.; Rosch, A.; Neubauer, A.; Georgii, R.; Böni, P. Skyrmion lattice in a chiral magnet *Science* **2009**, *323*, 915.
8. Wilhelm, H.; Baenitz, M.; Schmidt, M.; Rößler, U.; Leonov, A.; Bogdanov, A. Precursor phenomena at the magnetic ordering of the cubic helimagnet FeGe. *Phys. Rev. Lett.* **2011**, *107*, 127203. [CrossRef]
9. Münzer, W.; Neubauer, A.; Adams, T.; Mühlbauer, S.; Franz, C.; Jonietz, F.; Georgii, R.; Böni, P.; Pedersen, B.; Schmidt, M.; et al. Skyrmion lattice in the doped semiconductor Fe$_{1-x}$Co$_x$Si. *Phys. Rev. B* **2010**, *81*, 041203. [CrossRef]
10. Yu, X.Z.; Kanazawa, N.; Onose, Y.; Kimoto, K.; Zhang, W.Z.; Ishiwata, S.; Matsui, Y.; Tokura, Y. Near room-temperature formation of a skyrmion crystal in thin-films of the helimagnet FeGe *Nat. Mater.* **2011**, *10*, 106.
11. Tonomura, A.; Yu, X.; Yanagisawa, K.; Matsuda, T.; Onose, Y.; Kanazawa, N.; Park, H.S.; Tokura, Y. Real-space observation of skyrmion lattice in helimagnet MnSi thin samples. *Nano Lett.* **2012**, *12*, 1673–1677. [CrossRef] [PubMed]
12. Heinze, S.; Von Bergmann, K.; Menzel, M.; Brede, J.; Kubetzka, A.; Wiesendanger, R.; Bihlmayer, G.; Blügel, S. Spontaneous atomic-scale magnetic skyrmion lattice in two dimensions. *Nat. Phys.* **2011**, *7*, 713. [CrossRef]
13. Romming, N.; Hanneken, C.; Menzel, M.; Bickel, J.E.; Wolter, B.; von Bergmann, K.; Kubetzka, A.; Wiesendanger, R. Writing and deleting single magnetic skyrmions. *Science* **2013**, *341*, 636–639. [CrossRef] [PubMed]
14. Romming, N.; Kubetzka, A.; Hanneken, C.; von Bergmann, K.; Wiesendanger, R. Field-dependent size and shape of single magnetic skyrmions. *Phys. Rev. Lett.* **2015**, *114*, 177203. [CrossRef] [PubMed]
15. Woo, S.; Litzius, K.; Krüger, B.; Im, M.Y.; Caretta, L.; Richter, K.; Mann, M.; Krone, A.; Reeve, R.M.; Weigand, M.; et al. Observation of room-temperature magnetic skyrmions and their current-driven dynamics in ultrathin metallic ferromagnets. *Nat. Mater.* **2016**, *15*, 501. [CrossRef] [PubMed]

16. Soumyanarayanan, A.; Raju, M.; Oyarce, A.G.; Tan, A.K.; Im, M.Y.; Petrović, A.P.; Ho, P.; Khoo, K.; Tran, M.; Gan, C.; et al. Tunable room-temperature magnetic skyrmions in Ir/Fe/Co/Pt multilayers. *Nat. Mater.* **2017**, *16*, 898. [CrossRef] [PubMed]

17. Siemens, A.; Zhang, Y.; Hagemeister, J.; Vedmedenko, E.; Wiesendanger, R. Minimal radius of magnetic skyrmions: Statics and dynamics. *New J. Phys.* **2016**, *18*, 045021. [CrossRef]

18. Büttner, F.; Lemesh, I.; Beach, G.S. Theory of isolated magnetic skyrmions: From fundamentals to room temperature applications. *Sci. Rep.* **2018**, *8*, 4464. [CrossRef] [PubMed]

19. Jiang, W.; Zhang, X.; Yu, G.; Zhang, W.; Wang, X.; Jungfleisch, M.B.; Pearson, J.E.; Cheng, X.; Heinonen, O.; Wang, K.L.; et al. Direct observation of the skyrmion Hall effect. *Nat. Phys.* **2017**, *13*, 162. [CrossRef]

20. Litzius, K.; Lemesh, I.; Krüger, B.; Bassirian, P.; Caretta, L.; Richter, K.; Büttner, F.; Sato, K.; Tretiakov, O.A.; Förster, J.; et al. Skyrmion Hall effect revealed by direct time-resolved X-ray microscopy. *Nat. Phys.* **2017**, *13*, 170. [CrossRef]

21. Tomasello, R.; Martinez, E.; Zivieri, R.; Torres, L.; Carpentieri, M.; Finocchio, G. A strategy for the design of skyrmion racetrack memories. *Sci. Rep.* **2014**, *4*, 6784. [CrossRef] [PubMed]

22. Harris, V.G.; Pokhil, T. Selective-resputtering-induced perpendicular magnetic anisotropy in amorphous TbFe films. *Phys. Rev. Lett.* **2001**, *87*, 067207. [CrossRef] [PubMed]

23. Hansen, P.; Clausen, C.; Much, G.; Rosenkranz, M.; Witter, K. Magnetic and magneto-optical properties of rare-earth transition-metal alloys containing Gd, Tb, Fe, Co. *J. Appl. Phys.* **1989**, *66*, 756–767. [CrossRef]

24. Woo, S.; Song, K.M.; Zhang, X.; Zhou, Y.; Ezawa, M.; Liu, X.; Finizio, S.; Raabe, J.; Lee, N.J.; Kim, S.I.; et al. Current-driven dynamics and inhibition of the skyrmion Hall effect of ferrimagnetic skyrmions in GdFeCo films. *Nat. Commun.* **2018**, *9*, 959. [CrossRef] [PubMed]

25. Lee, J.T.; Chess, J.; Montoya, S.; Shi, X.; Tamura, N.; Mishra, S.; Fischer, P.; McMorran, B.; Sinha, S.; Fullerton, E.; et al. Synthesizing skyrmion bound pairs in Fe-Gd thin films. *Appl. Phys. Lett.* **2016**, *109*, 022402. [CrossRef]

26. Giannozzi, P.; Baroni, S.; Bonini, N.; Calandra, M.; Car, R.; Cavazzoni, C.; Ceresoli, D.; Chiarotti, G.L.; Cococcioni, M.; Dabo, I.; et al. QUANTUM ESPRESSO: A modular and open-source software project for quantum simulations of materials. *J. Phys. Cond. Matter* **2009**, *21*, 395502. [CrossRef]

27. Wimmer, E.; Krakauer, H.; Weinert, M.; Freeman, A.J. Full-potential self-consistent linearized-augmented-plane-wave method for calculating the electronic structure of molecules and surfaces: O_2 molecule *Phys. Rev. B* **1981**, *24*, 864. [CrossRef]

28. Kurz, P.; Förster, F.; Nordström, L.; Bihlmayer, G.; Blügel, S. Ab initio treatment of noncollinear magnets with the full-potential linearized augmented plane wave method. *Phys. Rev. B* **2004**, *69*, 024415. [CrossRef]

29. Eriksson, O.; Bergman, A.; Bergqvist, L.; Hellsvik, J. *Atomistic Spin Dynamics: Foundations and Applications*; Oxford University Press: Oxford, UK, 2017.

30. Antropov, V.P.; Katsnelson, M.I.; Harmon, B.N.; van Schilfgaarde, M.; Kusnezov, D. Spin dynamics in magnets: Equation of motion and finite temperature effects *Phys. Rev. B* **1996**, *54*, 1019. [CrossRef]

31. Katsnelson, M.I.; Irkhin, V.Y.; Chioncel, L.; Lichtenstein, A.I.; de Groot, R.A. Half-metallic ferromagnets: From band structure to many-body effects *Rev. Mod. Phys.* **2008**, *80*, 315. [CrossRef]

32. Vansteenkiste, A.; Leliaert, J.; Dvornik, M.; Helsen, M.; Garcia-Sanchez, F.; van Waeyenberge, B. The design and verification of MuMax3 *AIP Adv.* **2014**, *4*, 107133. [CrossRef]

33. Dzyaloshinskii, I.E. IE Dzyaloshinskii *Sov. Phys. JETP* **1957**, *5*, 1259.

34. Moriya, T. New mechanism of anisotropic superexchange interaction. *Phys. Rev. Lett.* **1960**, *4*, 228. [CrossRef]

35. Bode, M.; Heide, M.; Von Bergmann, K.; Ferriani, P.; Heinze, S.; Bihlmayer, G.; Kubetzka, A.; Pietzsch, O.; Blügel, S.; Wiesendanger, R. Chiral magnetic order at surfaces driven by inversion asymmetry. *Nature* **2007**, *447*, 190. [CrossRef] [PubMed]

applied sciences

MDPI

Article

Half-Metallicity and Magnetism of the Quaternary Heusler Compound TiZrCoIn$_{1-x}$Ge$_x$ from the First-Principles Calculations

Ying Chen [1,*]**, Shaobo Chen** [1]🆔**, Bin Wang** [2]**, Bo Wu** [3]🆔**, Haishen Huang** [3]**, Xinmao Qin** [1]**, Dongxiang Li** [1] **and Wanjun Yan** [1]

[1] School of Mathematics and Physics, Anshun College, Anshun 561000, China; shaobochen@yeah.net (S.C.); qxm200711@162.com (X.Q.); ldx0601@163.com (D.L.); yanwanjun7817@163.com (W.Y.)
[2] Anshun first senior high school, Anshun 561000, China; wangcowley@163.com
[3] Department of Physics, Zunyi Normal College, Zunyi 563002, China; fqwubo@163.com (B.W.); hhs_vir@yeah.net (H.H.)
* Correspondence: ychenjz@163.com

Received: 16 January 2019; Accepted: 8 February 2019; Published: 13 February 2019

Abstract: The effects of doping on the electronic and magnetic properties of the quaternary Heusler alloy TiZrCoIn were investigated by first-principles calculations. Results showed that the appearance of half-metallicity and negative formation energies are associated in all of the TiZrCoIn$_{1-x}$Ge$_x$ compounds, indicating that Ge doping at Z-site increases the stability without damaging the half-metallicity of the compounds. Formation energy gradually decreased with doping concentration, and the width of the spin-down gap increased with a change in Fermi level. TiZrCoIn$_{0.25}$Ge$_{0.75}$ was found to be the most stable half-metal. Its Fermi level was in the middle of the broadened gap, and a peak at the Fermi level was detected in the spin majority channel of the compound. The large gaps of the compounds were primarily dominated by the intense d-d hybridization between Ti, Zr, and Co. The substitution of In by Ge increased the number of sp valence electrons in the system and thereby enhanced RKKY exchange interaction and increased splitting. Moreover, the total spin magnetic moments of the doped compounds followed the Slater–Pauling rule of $M_t = Z_t - 18$ and increased from 2 μ_B to 3 μ_B linearly with concentration.

Keywords: quaternary Heusler alloy; doping; spin polarization; half-metallicity; magnetism

---

1. Introduction

Half-metallic ferromagnets [1–4] are potential candidates for spintronic applications owing to their completely spin-polarized band structures [5–8]. Of all known materials exhibiting half-metallicity, Heusler alloys have attracted considerable interest because of their high spin magnetic moments and high Curie temperatures [9–11]. Original and stoichiometric full-Heusler alloys have X$_2$YZ in their chemical formulas, where X and Y denote transition metal elements and Z is a primary group element. These Heuslers usually crystalize with a cubic L2$_1$ structure (space group Fm-3m, No. 255) and individually contain four interpenetrating fcc sublattices in their ideal forms. Several theoretical and experimental studies confirmed a series of Heusler compounds with half-metallicity [12–15]. Recently, quaternary Heusler alloys have increasingly attracted interest [16–21] owing to their half-metallicity and spin gapless band structures. These compounds show a distinct structural symmetry (space group F-43m, No. 216) resulting from the partial replacement of X in X$_2$YZ by another element X′. This crystal structure can be defined as a LiMgPdSn prototype. Most of these compounds, such as in Co$_2$Mn$_{1-x}$Fe$_x$Si, were designed for adjusting the Fermi energy to the middle of the gap [22]. Equivalent stoichiometric quaternary Heusler alloys have been extensively researched. CoFeMnZ compounds

(Z=Al, Ga, Si, or Ge) with 1: 1: 1: 1 stoichiometries show half-metallicity [23]. Özdoğan et al. performed a theoretical study on 60 LiMgPdSn-type quaternary Heusler alloys and found that most of the alloys exhibit half-metallicity [24]. Ideal properties, such as stable half-metallicity, large spin magnetic moment, and high Curie temperature, can be observed in the ordered bulk phase. Unfortunately, temperature, impurity, and other external factors can disrupt the completely spin-polarized band structures and further degrade the half-metallicity of ideal compounds [25–34]. For example, in the traditional full Heusler compound Co_2MnSi, a remarkable gap with 0.4 eV in minority band at 985 K may disappear, and only a spin polarization of 61% can be observed at a barrier interface consisting of a single electrode made of a Co_2MnSi film [35–38].

A dramatically large spin-down gap of 0.93 eV was detected in a quaternary Heusler TiZrCoIn [39]. The Ti of this compound was inclined to sit at A(0, 0, 0), Zr at B(0.25, 0.25, 0.25), Co at C(0.5, 0.5, 0.5), and In at Z(0.75, 0.75, 0.75) in the Wyckoff position coordinate. This compound might be an excellent candidate for spin-injectors because of its extraordinary wide band gap. Its performance for spintronic applications can be enhanced by modulating its magnetic and electronic properties on the basis of valence-electron count. In the present work, we searched for a mixed compound in the series $TiZrCoIn_{1-x}Ge_x$ where the half-metallic behavior is stable against the variation of impurity.

2. Calculation Methods

First-principles calculations were performed with the CASTEP code on the basis of the density functional theory (DFT). The stable ground structure of quaternary Heusler alloy TiZrCoIn was obtained by optimizing the structure. On the basis of the stable ground structure, one, two, three, or four Ge atoms were used to replace In atoms in a unit cell. Doping concentrations of 25%, 50%, 75%, and 100% were obtained. For the simulation of small doping concentrations of 12.5%, 6.25%, and 3.125%, TiZrCoIn supercells with 32, 64, and 128 atoms, respectively, were produced. The In atom was replaced by one Ge atom. Then, the correlation property calculations, such as the single point energy, density of states (DOS), and band-structure, were performed on the minimized doped cases obtained from the geometry optimization. In our calculations, all the electronic structure calculations were performed with the spin polarization. Generalized gradient approximation parameterized by Perdew [40,41] were used for the handling of the exchange and correlation term. Electron–ion interactions were disposed with ultrasoft pseudopotentials [42]. In the SCF calculation, a refined 11×10^{-6} eV/atom was used as a convergence criterion. For the energy cutoff of the planewave, the basis was set to 310 eV. The calculation was considered converged when the largest gradient was less than 0.002 eV/Å. In the property calculations, the $7 \times 7 \times 7$ special k-mesh points in Brillouin zone were applied. The electrons of Ti $3d^24s^2$, Zr $4d^25s^2$, Co $3d^74s^2$, In $5s^25p^1$, and Ge $4s^24p^2$ were regarded as valence electrons.

3. Results and Discussion

3.1. Formation Energy of Doped TiZrCoIn

The formation energy of the doped systems were analyzed with respect to the ordered case and estimated by using formula

$$E_f(dop) = E(dop) - E(TiZrCoIn) - \sum m_i\mu_i(bulk) \qquad (1)$$

where $E(dop)$ is the total energy of the doped case, and $E(TiZrCoIn)$ is the total energy of the ordered TiZrCoIn structure in the same supercell size. The last term $\sum m_i\mu_i(bulk)$ can be derived from the number of atoms added or removed from the ideal alloy. The number of altered atoms relative to that in the ideal bulk is represented by m_i, which is either positive, when atoms are added to the original system, or negative, when atoms are removed from system. The symbol $\mu_i(bulk)$ stands for the chemical potential of the corresponding atom in its bulk phase of ground state. Experimental preparation conditions and annealing environment can affect the formation of these doped systems. Thus, the formation energy of a doped structure is influenced by the chemical potential of the host atoms, which are affected by the environment. To obtain the chemical potentials from their bulk phase, we assumed that Ti, Zr, Co, In, and Ge are in the thermodynamic equilibrium with their bulk solid phase in the host rich condition. The hcp, hcp, hcp, fcc, and fcc structures for the bulk phases of Ti, Zr, Co, In, and Ge, respectively, were adopted.

The formation energy of each doped system was calculated with the formula above. The results are shown in the first column of Table 1. All the doped systems had negative formation energies. Thus, In atoms in the ordered phase were easily replaced by Ge atoms. The ideal bulk of TiZrCoIn alloy was affected by the impurity Ge during growth. The variations among these values indicated differences in stability. Lower formation energy indicates that the doped one is easier to be formed spontaneously during the growth. It is shown in table 1 the formation energy gets smaller with the doping concentration, which manifests the doped system tends to be formed at a high doping concentration. The highest formation energy was obtained at a doping concentration of 3.125%. However, it can still form spontaneously during growth owing to its negative formation energy (−9.256 eV). The lowest formation energy was obtained at 100% doping concentration. Hence, the doped compound (TiZrCoGe) is most likely to be fabricated experimentally compared to the other Ge doped compounds. Meanwhile, Ge doping at Z-site may further stabilize the compound.

Table 1. Formation energies (E_f), lattice parameters (unit in Å), spin polarizations presented as spin-up over spin-down rations, width of spin-down gap around the Fermi level (in units of eV), and total magnetic moments in a unit cell. Notice that x is the doping concentration.

x	E_f (eV)	a (Å)	$P(\frac{\uparrow-\downarrow}{\uparrow+\downarrow})$(%)	*Band Gap* (eV)	M_t (μ_B/f.u.)
x = 0	-	6.562	100	0.914	1.99
x = 3.125%	−9.256	6.633	100	0.681	2.03
x = 6.25%	−9.280	6.630	100	0.692	2.06
x = 12.5%	−9.333	6.609	100	0.721	2.13
x = 25%	−9.435	6.566	100	0.802	2.26
x = 50%	−9.604	6.496	100	0.904	2.50
x = 75%	−9.776	6.427	100	0.960	2.75
x = 100%	−10.109	6.356	100	1.130	3.00

3.2. Electronic Structure: Magnetic Moments and DOS

The quaternary Heusler compound TiZrCoIn with the configuration mentioned above showed half-metallic property and a broad indirect spin-down gap of 0.914 eV at the equilibrium lattice constant of 6.562 Å. This result fits perfectly with the previous prediction in Reference [39]. Lattice parameters, spin polarizations, width of spin-down gap, and total magnetic moments in a unit cell are listed in Table 1. The spin polarizations at the Fermi levels of all the doped structures were 100%. Thus, all the derived compounds were half-metal. In addition, the magnetic moments increased from 2 μ_B to 3 μ_B at increased doping concentration.

Before the detailed analysis of doping functional mechanism in quaternary Heusler TiZrCoIn, we initially discuss the ideal case of the compound TiZrCoIn. The spin-polarization total DOS and the atom PDOS are presented in the first line of Figures 1 and 2, respectively. In Figure 1, the spin-up band (shadowed area) of compound TiZrCoIn with ordered structure was metallic. Meanwhile, a wide energy gap was observed in the spin-down band at the Fermi level. The electrons at the Fermi level were totally polarized, and this condition resulted in 100% spin polarization. The width of the band gap was 0.914 eV, and the Fermi level was slightly near the bottom of the band gap. Thus, the half-metallicity of quaternary Heusler TiZrCoIn was susceptible to external influences. Meanwhile, the gap was primarily determined by the covalent hybridization between Ti and Co, as shown in Figure 2. The d states of Ti with low valence occupied the high energy area above the Fermi level, thereby forming the bonding bands. By contrast, the d states of Co with high valence were located at the low energy area below the Fermi level, thereby forming the antibonding bands. Ti and Co atoms have the same symmetry as in typical L2$_1$ full-Heuslers, and thus their d orbitals hybridize initially and form five bonding d hybrids and five nonbonding ones. Then, the five bonding d hybridized orbitals of Ti-Co hybridize with the d orbitals of the Zr atoms, thereby regenerating bonding and antibonding states. Given the large energy separation of d hybrids of the Ti and Co atoms, the five nonbonding d hybrids Ti-Co, namely, the t$_{1u}$ and e$_u$ states, possess high energies and are unoccupied. A detailed d-d hybridization diagram is shown in Figure 3. Introduced by the sp element In, a single degenerate s band and a triple degenerated p band, lying deep in energy, are located below the d states and accommodate d charge from the transition metal atoms. Thus, 9 instead of 12 occupied states are usually observed in the spin-down band. In the quaternary Huesler alloy TiZrCoIn, the In atom carried three valence electrons, and Ti, Zr, and Co carried four, four, and nine, respectively. Therefore, 17 transition metal electrons were found, of which 5 were cached by the s and p bands of the In atom, 10 were filled in the bonding d bands, and 2 were uncompensated. Thus, a total spin magnetic moment of 2 μ_B was observed in the compound TiZrCoIn. The total magnetic moments followed the Slater–Pauling rule: $M_t = Z_t - 18$, where M_t represents the total spin magnetic moments in a unit cell, and Z_t is the number of the total valence electron.

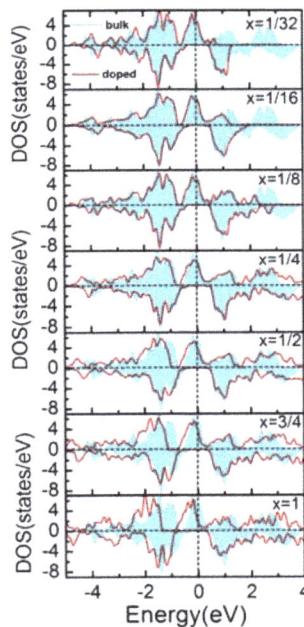

Figure 1. Total density of states (DOS) of the ordered TiZrCoIn quaternary Heusler alloy (shadowed area) and the doped systems (solid line). Note that all DOS are transited to a primitive cell. Notice that x in the figure is the doping concentration.

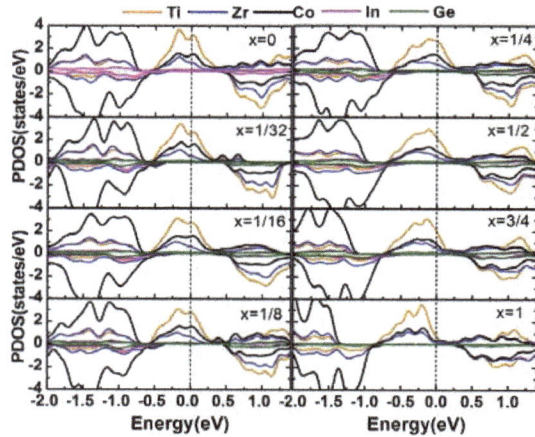

Figure 2. Partial density of states (PDOS) for the ordered TiZrCoIn quaternary Heusler alloy and the doped systems. The orange line is the atomic PDOS for Ti atom, the blue line is for the Zr atom, the black line is for the Co atom, the magenta line is for the In atom, and the olive line is for the Ge atom. Notice that x in the figure is the doping concentration.

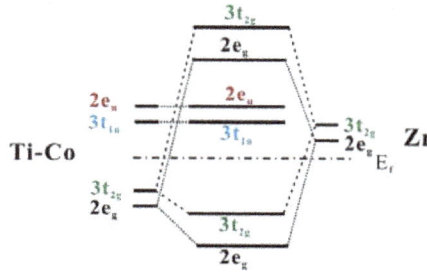

Figure 3. Schematic of the d-d hybridization of the ordered TiZrCoIn quaternary Heusler alloy.

Next, we will discuss the mixed compounds in the series TiZrCoIn$_{1-x}$Ge$_x$ in detail, to look for the compounds with better half-metallic stability. As shown in Figure 1, all the doped systems showed half-metallicity. The bonding states moved slightly to the low energy area. The higher the doping concentration is, the more pronounced the movement and the wider the band gaps are. The band gaps further widened from 0.914 eV to 1.13 eV at doping concentrations of 3.125%–100%. However, the Fermi levels shifted slightly from the middle of the gap to the edge of the gap. The bonding states in the spin-down channel of doped cases primarily originated from the transition metal Co with high valence, as shown in Figure 2. Meanwhile, the antibonding d states primarily came from the transition metal Ti with low valence. The d orbitals of Ti and Co atoms hybridized initially because of their identical symmetries. Then, the creating bonding d states in turn hybridized with the d orbitals of the Zr atoms. The outlines of the d states near the Fermi levels of Ti, Zr, and Co were nearly identical, especially those of Ti and Co. Therefore, Ti, Co, and Zr had intense interactions. After the addition of the doped element Ge, the number of sp valence electrons of the parent compound increased, and the increase indirectly enhanced d-d hybridization. The d-d hybridization dominated the half-metallicity of the compounds. Nonetheless, increase in total spin moment and energy band gap depend on the sp atoms. This dependence plays a substantial role in the RKKY exchange interaction.

4. Conclusions

The effects of doping on the electronic and magnetic properties of the quaternary Heusler alloy TiZrCoIn with half-metallicity were investigated with first-principles calculations based on DFT. The results showed that all the doped compounds carry negative formation energies and thus can be formed during the growth of the TiZrCoIn compound. The formation energy decreased with doping concentration.It was found that the formation energy was at its lowest value (-1.13 eV) when x=1. Namely, the quaternary Heusler alloy TiZrCoGe derived from doping showed stable half-metallicity. In all the cases, a wide spin-minority band gap was observed at the Fermi level, and the gap widened as the doping concentration increased. The Fermi level was located in the middle of the broadened gap, and a peak at the Fermi level was detected in the spin majority channel when x was 0.75. Thus, the compound $TiZrCoIn_{0.25}Ge_{0.75}$ showed high half-metallic stability. The total magnetic moments of the doped compounds followed the $M_t = Z_t - 18$ SP rule, increasing from 2 μ_B to 3 μ_B linearly with doping concentration. The d-d hybridization among Ti, Zr, and Co dominated the half-metallicity, and sp atoms (In and Ge) played an important role in the RKKY exchange interaction.

Author Contributions: Methodology, Y.C.; software, W.Y. and B.W. (Bo Wu); formal analysis, B.W. (Bin Wang) and S.C.; investigation, Y.C. and D.L.; data curation, H.H. and X.Q.; writing—original draft preparation, S.C. and B.W. (Bin Wang); writing—review and editing, Y.C. and B.W. (Bo Wu)

Funding: This research was funded by the key projects of the tripartite foundation of the Guizhou Science and Technology Department (Grant No. [2017]7041), the Natural science research project of Guizhou provincial department of education (Grant No. [2017]288), the Key Projects of the Tripartite Foundation of Guizhou Science and Technology Department (Grant No. [2015]7696), and the project of the creative research groups of Guizhou Province of China (Grant No. [2016]048).

Acknowledgments: We are really grateful to the cloud computing platform at Guizhou University and the CNROCK HOLE BLACKHOLE high density computing platform for computing support.

References

1. Ohnuma, Y.; Matsuo, M.; Maekawa, S. Spin transport in half-metallic ferromagnets. *Phys. Rev. B.* **2016**, *94*, 184405. [CrossRef]
2. Griffin, S.M.; Neaton, J.B. Prediction of a new class of half-metallic ferromagnets from first principles. *Phys. Rev. Mater.* **2017**, *1*. [CrossRef]
3. Kourov, N.I.; Marchenkov, V.V.; Belozerova, K.A.; Weber, H.W. Specific features of the electrical resistivity of half-metallic ferromagnets Fe₂MeAl (Me = Ti, V, Cr, Mn, Fe, Ni). *J. Exper. Theor. Phys.* **2014**, *118*, 426. [CrossRef]
4. Sun, M.; Ren, Q.; Zhao, Y.; Wang, S.; Yu, J. Magnetism in transition metal-subsitiued germanane: A search for room temperature spintronic devices. *J. Appl. Phys.* **2016**, *119*, 666. [CrossRef]
5. Li, S.; Takahashi, Y.K.; Sakuraba, Y.; Chen, J.; Furubayashi, T.; Mryasov, O.; Faleev, S.; Hono, K. Current-perpendicular-to-plane giant magnetoresistive properties in Co₂Mn(Ge₀.₇₅Ga₀.₂₅)/Cu₂TiAl/Co₂Mn(Ge₀.₇₅Ga₀.₂₅) all-Heusler alloy pseudo spin valve. *J. Appl. Phys.* **2016**, *119*, 093911. [CrossRef]
6. Galanakis, I.; Özdoğan, K.; Şaşioğlu, E. Spin-filter and spin-gapless semiconductors: The case of Heusler compounds. *AIP Adv.* **2016**, *6*, 055606. [CrossRef]
7. Wang, Y.; Ramaswamy, R.; Yang, H. FMR-related phenomena in spintronic devices. *J. Phys. D. Appl. Phys.* **2018**, *51*, 27. [CrossRef]
8. Feng, Y.; Cui, Z.; Wei, M.S.; Wu, B. Spin-polarized quantum transport in Fe4N based current-perpendicular-to-plane spin valve. *App. Surf. Sci.* **2019**, *499*, 78–83. [CrossRef]
9. Wang, X.T.; Cheng, Z.X.; Wang, W.H. L2₁ and XA Ordering Competition in Hafnium-Based Full-Heusler Alloys Hf₂VZ (Z = Al, Ga, In, Tl, Si, Ge, Sn, Pb). *Materials* **2017**, *10*, 1200. [CrossRef]
10. Gregor, F.; Perter, K. Ternary semiconductors NiZrSn and CoZrBi with half-Heusler structure: A first-principles study. *Phys. Rev. B.* **2016**, *94*, 075203.

11. Barman, C.K.; Alam, A. Topological phase transition in the ternary half-Heusler alloy ZrIrBi. *Phys. Rev. B.* **2018**, *97*, 075302. [CrossRef]

12. Galanakis, I. Theory of Heusler and Full-Heusler Compounds. *Springer Series in Materials Science.* **2016**, *222*, 3–36.

13. Palmstrom, C.J. Heusler compounds and spintronics. *Prog. Crys. Ggro. Chara. Mater.* **2016**, *2*, 371. [CrossRef]

14. Oliynyk, A.O.; Antono, E. High-Throughput Machine-Learning-Driven Synthesis of Full-Heusler Compounds. *Chem. Mater.* **2016**, *28*, 20. [CrossRef]

15. Anjami, A.; Boochani, A.; Elahi, S.M.; Akbari, H. Ab-initio study of mechanical, Half-metallic and optical properties of Mn_2ZrX (X = Ge, Si) compounds. *Results Phys.* **2017**, *7*, 3522–3529. [CrossRef]

16. Ashis, K.; Srikrishna, G.; Rudra, B.; Subbhradip, G.; Biplab, S. New quaternary half-metallic ferromagnets with large Curie temperatures. *Sci. Reports* **2017**, *7*, 1803.

17. Jafari, K.; Ahmadian, F. First-principles Study of Magnetism and Half-Metallic Properties for the Quaternary Heusler Alloys CoRhYZ (Y = Sc, Ti, Cr, and Mn; Z = Al, Si, and P). *J. Super. Novel. Magn.* **2017**, *14*, 1–10. [CrossRef]

18. Bahramian, S.; Ahmadian, F. Half-metallicity and magnetism of quaternary Heusler compounds CoRuTiZ(Z = Si, Ge, and Sn). *J. Magn. Magn. Mater.* **2017**, *424*, 122–129. [CrossRef]

19. Rani, D.; Suresh, E.K.G.; Yadav, A.K.; Jha, S.N.; Varma, D.M.R.; Alam, A. Structural, electronic, magnetic, and transport properties of the equiatomic quaternary Heusler alloy CoRhMnGe: Theory and experiment. *Phys. Rev. B.* **2017**, *96*, 184404. [CrossRef]

20. Wang, X.T.; Cheng, Z.X.; Liu, G.D. Rare earth-based quaternary Heusler compounds MCoVZ (M = Lu, Y; Z = Si, Ge) with tunable band characteristics for potential spintronic applications. *IUCrJ* **2017**, *4*, 758–768. [CrossRef]

21. Li, Y.P.; Liu, G.D.; Wang, X.T.; Liu, E.K. First-principles study on electronic structure, magnetism and half-metallicity of the NbCoCrAl and NbRhCrAl compounds. *Results Phys.* **2017**, *7*, 2248–2254. [CrossRef]

22. Kallmayer, M.; Elmers, H.J.; Balke, B.; Wurmehl, S.; Emmerling, F. Magnetic properties of $Co_2Mn_{1-x}Fe_xSi$ Heusler alloys. *J. Phys. D: Appl. Phys.* **2006**, *39*, 786. [CrossRef]

23. Alijani, V.; Ouardi, S.; Fecher, G.H.; Jürgen, W.; Naghavi, S.S.; Kozina, X.; Stryganyuk, G.; Felser, C.; Ikenaga, E.; Yamashita, Y.; et al. Electronic, structural, and magnetic properties of the half-metallic ferromagnetic quaternary Heusler compounds CoFeMnZ (Z = Al, Ga, Si, Ge). *Phys. Rev. B.* **2011**, *73*, 2022–2399.

24. Özdoğan, K.; Şaşioğlu, E.; Galanakis, I. Slater–Pauling behavior in LiMgPdSn-type multifunctional quaternary Heusler materials: Half-metallicity, spin-gapless and magnetic semiconductors. *J. Appl. Phys.* **2013**, *113*, 323-R. [CrossRef]

25. Feng, Y.; Chen, H.; Yuan, H.; Zhou, Y.; Chen, X. The effect of disorder on electronic and magnetic properties of quaternary Heusler alloy CoFeMnSi with LiMgPbSb-type structure. *J. Magn. Magn. Mater.* **2015**, *378*, 7–15. [CrossRef]

26. Yang, G.; Li, D.; Wang, S.; Ma, Q.; Liang, S.; Wei, H.; Han, X.; Hesjedal, T.; Ward, R.; Kohn, A. Effect of interfacial structures on spin dependent tunneling in epitaxial L10-FePt/MgO/FePt perpendicular magnetic tunnel junctions. *J. Appl. Phys.* **2015**, *117*, 083904. [CrossRef]

27. Feng, Y.; Xu, X.; Cao, W.; Zhou, T. Investigation of cobalt and silicon co-doping in quaternary Heusler alloy NiFeMnSn. *Comp. Mater. Sci.* **2018**, *147*, 251–257. [CrossRef]

28. Han, J.; Gao, G. Large tunnel magnetoresistance and temperature-driven spin filtering effect based on the compensated ferrimagnetic spin gapless semiconductor Ti_2MnAl. *Journal Title* **2018**, *113*, 102402. [CrossRef]

29. Aron-Dine, S.; Pomrehn, G.S.; Pribram-Jones, A.; Laws, K.J.; Bassman, L. First-principles investigation of structural and magnetic disorder in CuNiMnAl and CuNiMnSn Heusler alloys. *Phys. Rev. B.* **2017**, *95*, 024108. [CrossRef]

30. Hu, Y.; Zhang, J.M. First-principles study on the thermodynamic stability, magnetism, and half-metallicity of full-Heusler alloy Ti_2FeGe (001) surface. *Phys. Lett. A.* **2017**, *381*, 1592–1597. [CrossRef]

31. Feng, Y.; Wu, B.; Yuan, H.; Chen, H. Structural, electronic and magnetic properties of $Co_2MnSi/Ag(100)$ interface. *J. Alloy. Compd.* **2015**, *623*, 29–35. [CrossRef]

32. Feng, Y.; Chen, X.; Zhou, T.; Yuan, H.; Chen, H. Structural stability, half-metallicity and magnetism of the CoFeMnSi/GaAs(001) interface. *Appl. Surf. Sci.* **2015**, *346*, 1–10. [CrossRef]

33. Zhang, S.; Jin, Z.; Liu, X.; Zhao, W.; Lin, X.; Jing, C.; Ma, G. Photoinduced terahertz radiation and negative conductivity dynamics in Heusler alloy Co_2MnSn film. *Opt. Lett.* **2017**, *42*, 3080. [CrossRef]

34. Chen, Y.; Wu, B.; Yuan, H.K.; Feng, Y.; Chen, H. The defect-induced changes of the electronic and magnetic properties in the inverse Heusler alloy Ti_2CoAl. *J. Solid. State. Chem.* **2015**, *221*, 311. [CrossRef]

35. Branford, W.R.; Singh, L.J.; Barber, Z.H.; Kohn, A.; Petfordlong, A.K. Temperature insensitivity of the spin-polarization in Co_2MnSi films on GaAs (001). *New. J. Phys.* **2016**, *9*, 224. [CrossRef]

36. Pradines, B.; Arras, R.; Abdallah, I.; Biziere, N.; Calmels, L. First-principles calculation of the effects of partial alloy disorder on the static and dynamic magnetic properties of Co_2MnSi. *Phys. Rev. B.* **2017**, *95*, 094425. [CrossRef]

37. Hu, B.; Moges, K.; Honda, Y.; Liu, H.X.; Uemura, T.; Yamamoto, M.; Inoue, J.I.; Shirai, M. Temperature dependence of spin-dependent tunneling conductance of magnetic tunnel junctions with half-metallic Co_2MnSi electrodes. *Phys. Rev. B.* **2016**, *94*, 094428. [CrossRef]

38. Andrieu, S.; Neggache, A.; Hauet, T.; Devolder, T.; Hallal, A.; Chshiev, M.; Bataille, A.M.; Fèvre, P.; Bertran, F. Direct evidence for minority spin gap in the Co_2MnSi Heulser compound. *Phys. Rev. B.* **2016**, *93*, 094417. [CrossRef]

39. Yan, P.L.; Zhang, J.M.; Xu, K.W. The structural, electronic and magnetic properties of quaternary Heusler alloy TiZrCoIn. *Soli. State. Commu.* **2016**, *231–232*, 64–67. [CrossRef]

40. Perdew, J.P.; Burke, K.; Ernzerhof, M. Generalized gradient approximation made simple. *Phys. Rev. Lett.* **1996**, *77*, 3865. [CrossRef]

41. Noemí, H.H.; Joaquín, O.C.; Martynov, Y.B.; Nazmitdinov, R.G.; Frontera, A. DFT prediction of band gap in organic-inorganic metal halide perovskites: An exchange-correlation functional benchmark study. *Chem. Phys.* **2019**, *516*, 225–231.

42. Garcia, A.; Junquera, J.; Portal, D.S.; Soler, J.M. Electronic Structure Calculations with Localized Orbitals: The Siesta Method. *Spri. Neth.* **2005**, 77–91.

applied
sciences

MDPI

Article

Electronic, Optical, Mechanical and Lattice Dynamical Properties of MgBi$_2$O$_6$: A First-Principles Study

Lin Liu [1], Dianhui Wang [1,2] , Yan Zhong [1,2] and Chaohao Hu [1,2,*]

[1] School of Materials Science and Engineering, Guilin University of Electronic Technology, Guilin 541004, China; liulinguet@163.com (L.L.); dhwang@guet.edu.cn (D.W.); yanzhong@guet.edu.cn (Y.Z.)
[2] Guangxi Key Laboratory of Information Materials, Guilin University of Electronic Technology, Guilin 541004, China
[*] Correspondence: chaohao.hu@guet.edu.cn

Received: 28 February 2019; Accepted: 18 March 2019; Published: 27 March 2019

Abstract: Electronic structure, optical, mechanical, and lattice dynamical properties of the tetragonal MgBi$_2$O$_6$ are studied using a first-principles method. The band gap of MgBi$_2$O$_6$ calculated from the PBE0 hybrid functional method is about 1.62 eV and agrees well with the experimental value. The calculations on elastic constants show that MgBi$_2$O$_6$ exhibits mechanical stability and strong elastic anisotropy. The detailed analysis of calculated optical parameters and effective masses clearly indicate that MgBi$_2$O$_6$ has strong optical response in the visible light region and high separation efficiency of photoinduced electrons and holes.

Keywords: MgBi$_2$O$_6$; optical properties; mechanical anisotropy; lattice dynamics; first-principles calculations

1. Introduction

Currently, Bi-based oxides have received considerable attention due to their particular physical properties and wide applications in different fields like multiferroics [1,2], superconductivity [3,4], and photocatalysis [5,6]. Generally, Bi exists as the trivalent state (Bi^{3+}) in most of the Bi-based oxides like Bi$_2$O$_3$ [7], BiVO$_4$ [8], Bi$_2$WO$_6$ [9], Bi$_2$Sn$_2$O$_7$ [10], BiFeO$_3$ [11], and BiMnO$_3$ [12]. However, some Bi-containing oxides with the unusual pentavalent state (Bi^{5+}) have also attracted research interest. For example, NaBiO$_3$ has been found to show the absorption of visible light and can be used as a prominent material for photooxidation of organics [13]. A recent work by Gong et al. [14] shows that AgBiO$_3$ can self-produce significant amounts of reactive oxygen species without light illumination or any other additional oxidant and has an excellent oxidizing reactivity. BaBiO$_3$, as one kind of Bi-based oxides containing Bi^{3+} and Bi^{5+} mixed valent states, has been found to show the potential use for the absorber of all-oxide photovoltaics [15] and can be an active photocatalyst under visible-light irradiation [16].

MgBi$_2$O$_6$ adopting the trirutile-type structure is also a Bi^{5+}-containing compound. Kumada et al. [17] first successfully prepared MgBi$_2$O$_6$ by the low-temperature hydrothermal method and characterized its crystal structure in detail. In 2003, Mizoguchi et al. [18] investigated the optical and electrical properties of MgBi$_2$O$_6$ and found that MgBi$_2$O$_6$ is a degenerate n-type semiconductor with the band gap of about 1.8 eV, and it is possible to produce an optical band gap that extends into the visible region of the spectrum by tuning the Bi^{5+} 6s–O 2p interaction in order to produce a disperse conduction band. The band structure makes MgBi$_2$O$_6$ a good candidate of visible light-sensitive photocatalysts for decomposition of organic species. Comparing with other pentavalent bismuthates such as LiBiO$_3$, NaBiO$_3$, KBiO$_3$, ZnBi$_2$O$_6$, SrBi$_2$O$_6$, AgBiO$_3$, BaBi$_2$O$_6$, and PbBi$_2$O$_6$, MgBi$_2$O$_6$ does not possess the highest photocatalytic activity, but there is no adsorption observed in the decomposition of methylene blue [19]. It may be a good idea to adjust the wide band gap of some traditional photocatalysts such TiO_2 by building complex compounds with MgBi2O6. In our recent work, the photocatalytic activity of $MgBi_2O_6$ has been found to be significantly enhanced via constructing AgBr/$MgBi_2O_6$ heterostructured composites [20]. Theoretically, the band gap of $MgBi_2O_6$ was calculated to be about 1.10 eV using the Heyd-Scuseria-Ernzerhof (HSE) functional method within the framework of the density functional theory (DFT), and was found to be widely tuned by applying external strain [21]. On the basis of theoretical calculations, however, the deeper understanding of the physical properties of $MgBi_2O_6$ is still in lacking. In this work, we investigated the electronic structure, optical, mechanical, and lattice dynamical properties $MgBi_2O_6$ using first-principles calculations.

2. Computational Details

All calculations were carried out using the Vienna ab-initio simulation package (VASP) [22,23], which is an implement of DFT. The projector augmented wave (PAW) method [24,25] was used to describe the ion-electron interactions. The generalized gradient approximation (GGA) parameterized by Perdew, Burke and Ernzerhof (PBE) [26] was applied for the exchange-correlation function. The cutoff energy for the plane wave basis set was fixed at 500 eV. The Brillouin zone is sampled by a Monkhorst-Pack type k-point mesh with density of $2\pi \times 0.03$ Å^{-1}. Full relaxation of $MgBi_2O_6$ unit cell was performed until the changes in total energy and force on each atom are less than 10^{-5} eV and 10^{-3} eV/Å, respectively. In this work, Mg-3s, Mg-2p, Bi-6s, Bi-6p, O-2s, and O-2p states are taken as valence electrons. Based on the DFT calculation, a more accurate screened coulomb hybrid functional (HF) developed by Heyd, Scuseria, and Ernzerhof (PBE0) [27,28] (containing 30% of the exact exchange, and 70% of the PBE exchange, and 100% of the PBE correlation energy in this work) was used to calculate the electronic structure. In the calculations of optical properties, a dense k-point density of $2\pi \times 0.015$ Å^{-1} including the Γ point was used to ensure the accurate precision. Moreover, 88 extra empty bands are included in the calculations to hold the excited electronic states. Phonon calculations were performed within the framework of density functional perturbation theory to investigate the lattice dynamical properties of $MgBi_2O_6$. The phonon dispersion curve and phonon density of states (PDOS) were calculated by the PHONOPY code [29] on the basis of force constants obtained from VASP code. In the phonon calculations, a $2 \times 2 \times 1$ supercell was used.

3. Results and Discussion

3.1. Optimized Crystal Structure of MgBi2O6

At ground state, $MgBi_2O_6$ crystallizes in a tetragonal trirutile-type structure with $P4_2/mnm$ space group. It can be seen from the crystal structure depicted in Figure 1 that BiO_6 and MgO_6 octahedrons connect with the sharing edges and stack along the [001] direction by 2:1 ratio. In the [110] and [$\bar{1}$10] directions, BiO_6 and MgO_6 octahedrons are connected with each other by sharing O vertices. The optimized lattice constants listed in Table 1 are in good agreement with the experimental values [17]. The BiO_6 octahedrons are slightly distorted and the bond lengths of Bi-O(8j) are not completely identical.

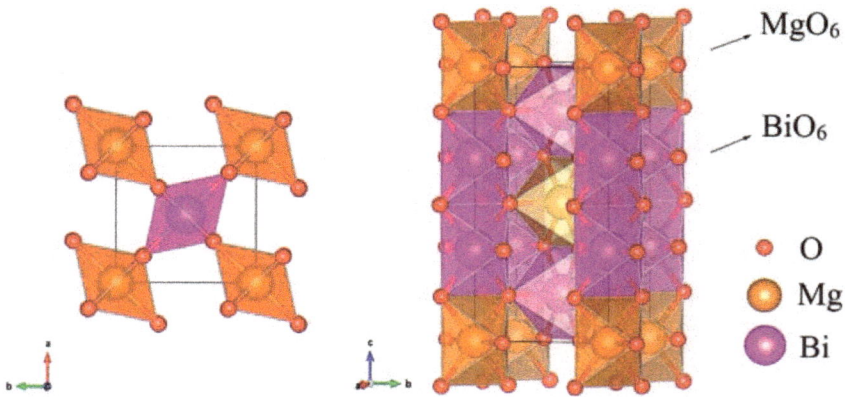

Figure 1. Crystal structure of the trirutile-type $MgBi_2O_6$ represented as two different views.

Table 1. Optimized structural parameters of $MgBi_2O_6$, Bi-O and Mg-O interatomic distances. The experimental values [17] are listed in brackets for comparison.

Lattice Constants (Å)	Atomic Coordinates (Fractional)				
	Atom	Site	x	y	z
$a = 4.920$ (4.826)	Bi	4e	0	0	0.333 (0.332)
$c = 9.924$ (9.719)	Mg	2a	0	0	0
	O1	4f	0.302 (0.305)	0.302 (0.305)	0
	O2	8j	0.309 (0.307)	0.309 (0.307)	0.336 (0.335)
BiO_6 octahedron			**MgO_6 octahedron**		
Bi-O(1)	2.156 (2.114) $\times 2$		Mg-O(1)	2.107 (2.084) $\times 2$	
Bi-O(2)	2.153 (2.102) $\times 2$		Mg-O(2)	2.103 (2.073) $\times 4$	
Bi-O(3)	2.141 (2.084) $\times 2$				

3.2. Electronic Properties of $MgBi_2O_6$

To study the electronic property of tetragonal $MgBi_2O_6$, the electronic band structure along high symmetry directions in the Brillouin zone (BZ) and density of states are calculated. The band structure calculated by conventional DFT (dotted line in Figure 2) shows that there is a bit of overlap between the conduction band and valence band, indicating the metallic feature of $MgBi_2O_6$. This contradicts the experimental findings. It is well known that DFT calculations do not take into account the effects of electron excitation and thus tend to underestimate the electronic band gap. To acquire the more accurate results, the calculations based on the HF PBE0 method are further performed. The HF corrected band structure (solid line in Figure 2) shows that $MgBi_2O_6$ is a direct semiconductor with a band gap of about 1.62 eV, which is well consistent with the experimental measured data (1.6~1.8 eV) [18–20] and larger than the previously calculated value (1.10 eV) based on the HF method [21]. The valence band near the Fermi level (E_F) is flat, while the dispersion of the conduction band close to the E_F is relative strong.

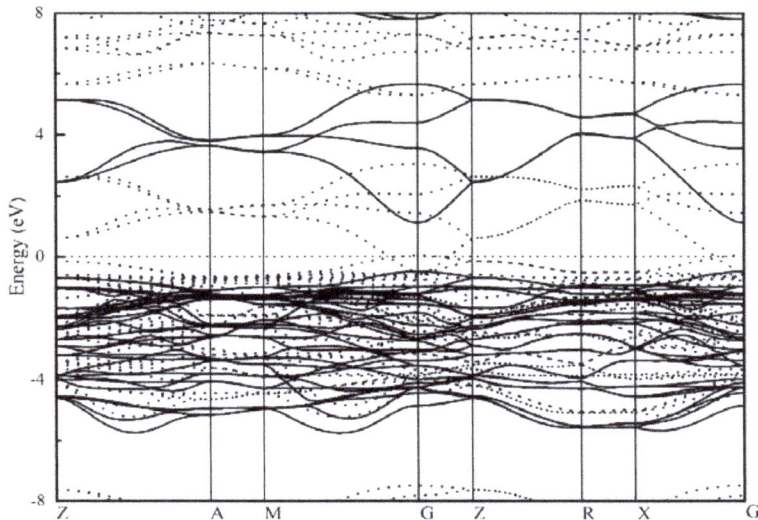

Figure 2. Calculated electronic band structure of MgBi$_2$O$_6$ using hybrid functional PBE0 method. Band structure from conventional DFT calculation (dotted line) is also presented for comparison.

The effective mass (m^*) of carriers is an important parameter, because it determines the transfer and separation efficiency of electrons and holes, and directly influences the photophysical properties of semiconductors. m^* can be calculated as follows:

$$\left(\frac{1}{m^*}\right)_{ij} = \frac{1}{\hbar^2}\frac{\partial^2 E(k)}{\partial k_i k_j}, \ i,j = x,\ y,\ z, \tag{1}$$

where $E_n(k)$ represents the band energy, k is the wave vector, and $\hbar$ is the reduced Planck constant. For MgBi$_2$O$_6$, the effective electron mass (m_e^*) at the conduction band and hole mass (m_h^*) at the valence band at the Γ point are calculated and listed in Table 2. The calculated tensors of m^* clearly show that m_e^* is fairly isotropic and m_h^* shows a relatively strong anisotropy. m_h^* along the [001] direction is larger than those along the [100] and [010] directions. The values of m_e^* in Table 2 are also comparable to the previously calculated value (0.277) for MgBi$_2$O$_6$ [21]. Moreover, the values of m_h^* are distinctly larger than those of m_e^*, which indicates that the mobility of holes at the valence band is obviously slower than that of electrons at the conduction band. The big difference in mobility between electrons and holes is undoubtedly beneficial to the separation of charge carriers and the reduction of the recombination rate of electron-hole pairs. In addition, the values of m_h^*/m_e^* of MgBi$_2$O$_6$ are in the range of 6.2~10.9, being larger than the corresponding value (2.1) of anatase TiO$_2$ [30] widely investigated as an important photocatalytic material. Thus, semiconducting MgBi$_2$O$_6$ can be considered as a photocatalyst with highly efficient separation of photoinduced electrons and holes.

Table 2. Calculated effective electron mass (m_e^*) and hole mass (m_h^*) along the three principal directions at the Γ point. All values are in units of free electron mass (m_0).

	Conduction Band			Valence Band		
Γ	0.221	0	0	−1.572	0	0
	0	0.206	0	0	−1.370	0
	0	0	0.171	0	0	−1.871

The calculated electronic total and partial density of states (DOS) of $MgBi_2O_6$ is shown in Figure 3. It can be clearly seen that the DOS near the top of the valence band is derived from the Bi-5*d*, Mg-2*p* and O-2*p* states. The bottom of the conduction band is mainly composed of O-2*p*, Bi-6*s*, and Mg-3*s* states. The strong hybridizations between Bi-5*d* and O-2*p* states in BiO_6 octahedrons and bonding interactions between Mg-2*p* or Mg-3*s* and O-2*p* states in MgO_6 octahedrons should be directly responsible for the structural stability of $MgBi_2O_6$.

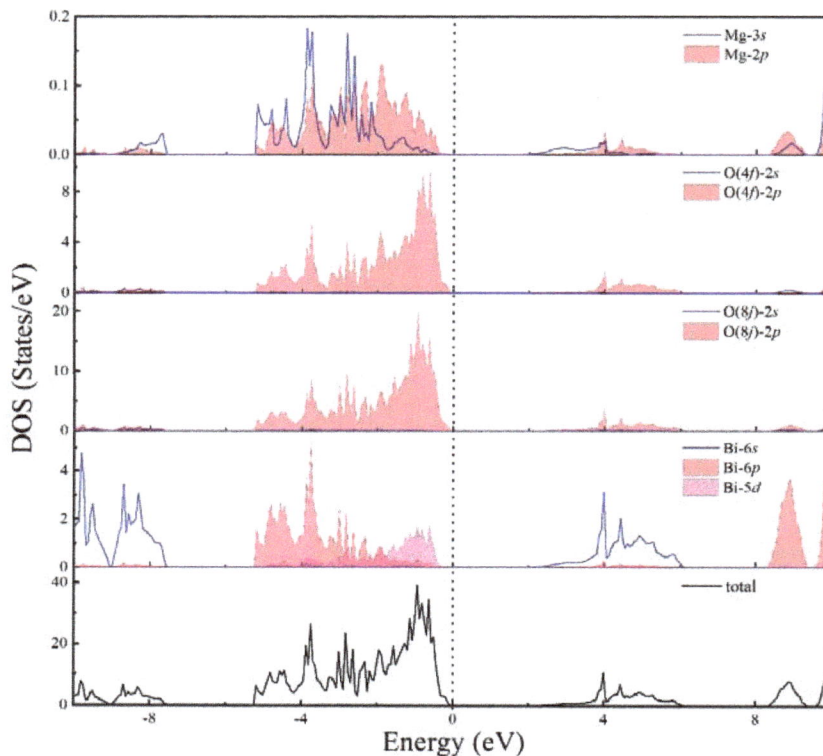

Figure 3. Calculated electronic density of states of $MgBi_2O_6$ using PBE0 method.

To get a deeper understanding of the bonding nature of Bi, Mg, and O atoms, the electron localization function (ELF) was further calculated. According to the original definition, the ELF values are scaled in the range from 0 to 1. The high ELF means strong covalent bonding interaction between atoms and the very low ELF close to 0 corresponds to the ionic bonding. Figure 4 shows the calculated ELF for (110) plane of $MgBi_2O_6$ passing through Bi, Mg, and O atoms. The ELF values between Mg and O atoms are extremely low, indicating the ionic bonding in MgO_6 octahedrons. This can be well understood on the basis of the previously obtained recognition on MgO, which is usually considered as one kind of classical ionically bonded compounds. The maximal value of ELF between Bi and O atoms reaches about 0.61, which clearly identifies the partially Bi-O covalent bonding interaction in BiO_6 octahedrons.

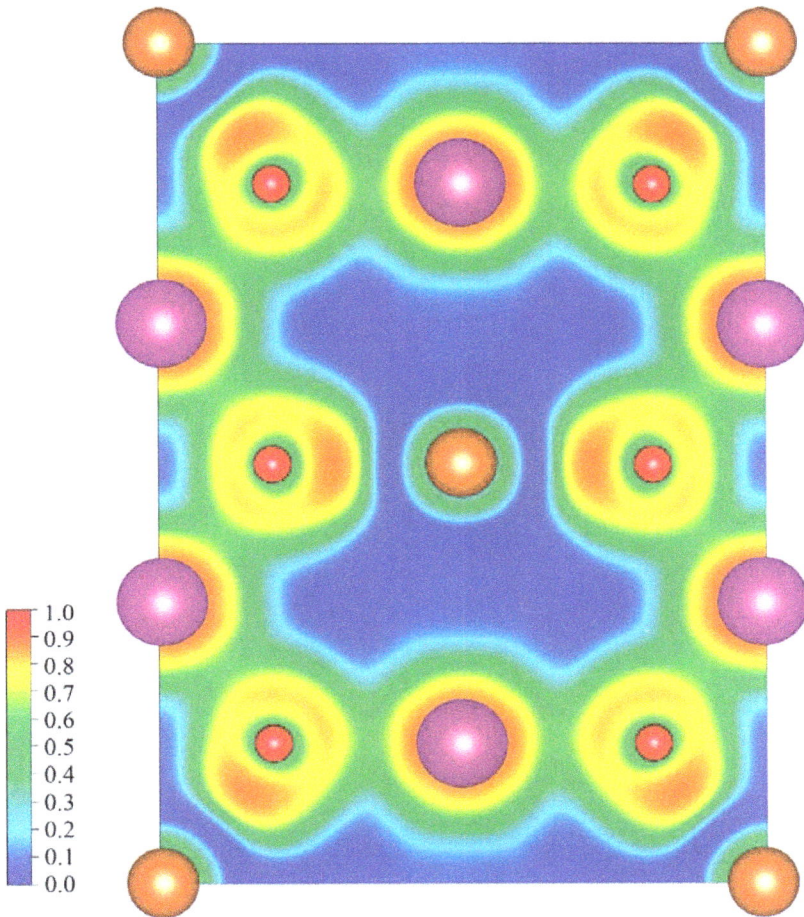

Figure 4. Calculated electron localization function for (110) plane across Bi, Mg, and O atoms.

3.3. Optical Properties of $MgBi_2O_6$

In order to understand the optical performance of semiconducting $MgBi_2O_6$, we calculated its dielectric functions using the hybrid functional PBE0 method, and further computed its optical properties such as complex refractive index n, extinction coefficient k, reflectivity R, absorption coefficients α, and electron energy-loss function L. The calculated dielectric functions are depicted in Figure 5. It can be clearly seen that the static dielectric function is found to be 4.1 at 0 eV. With the increase of the polarization intensity and energy, the dielectric function, the real part ε_1 is also gradually increased and reaches a maximum value of 5.4 when the energy value is about 2.8 eV. When the photon energy reaches 1.6 eV, which is the width of electronic band gap, the electrons in the valence bands begin to excite and transit to conduction bands. As the carrier concentration increases, the degree of polarization decreases and the dielectric function decreases slightly. With the further increment of photon energy, the dielectric function values begins to fluctuate, which corresponds to the change of carrier concentration in the crystal. The imaginary part ε_2 reflects the transition between occupied and non-occupied electrons and can be used to characterize the light absorption behavior of the crystal. The first two peaks of the imaginary part of the dielectric function ε_2 are, respectively, at about 2.0 and

3.0 eV, which probably results from the electron transition from the top of the valence bands to the bottom of the conduction bands.

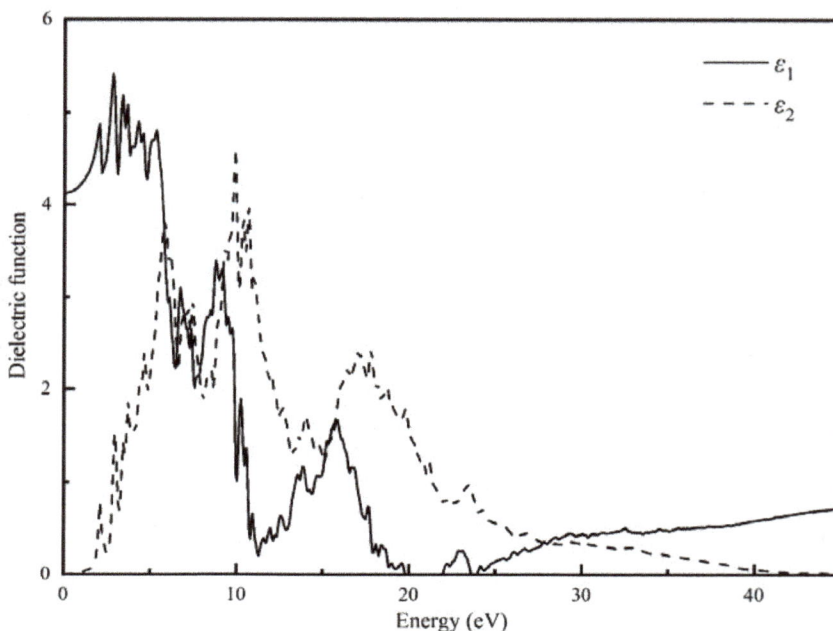

Figure 5. Calculated complex dielectric function of $MgBi_2O_6$.

The calculated refractive index n, extinction coefficient k, absorption coefficient α, reflectivity R, and electron energy loss function L of $MgBi_2O_6$ are shown in Figure 6. As presented in Figure 6a, the higher value of n is in the energy range of 1.5–5.8 eV and the corresponding wavelength range is from 214 to 826 nm, indicating that $MgBi_2O_6$ has a strong refractive effect in both ultraviolet (UV) and visible light region. The value of k rapidly starts to increase from 2.0 eV, also showing a response in the visible and UV light region. The calculated α in Figure 6c shows that the light absorption edge is about 1.6 eV and is comparable with the band gap calculated by the PBE0 method. The value of optical absorption edge also agrees with literatural values [18,20]. With the increase of energy, α also gradually increases and a series of absorption peaks appear in the energy range from 1.6 to 27 eV. Combining with the calculated dielectric function and density of states, we can find that the first two absorption peaks at 4.5 and 6.0 eV are probably related to the electron migration of the O-2p, Bi-6s, Bi-5d, and Mg-2p states. The reflectivity R and the electron energy loss function L can be used to represent the resonant frequency of the incident light and the resonant frequency of the plasma. As shown in Figure 6d, the average value of R is about 13.8%, indicating that $MgBi_2O_6$ can be used as a light absorbing material. The calculated L presented in Figure 6e completely locates in the continuous energy range of 0–45 eV, indicating that the characteristics of plasma oscillation in $MgBi_2O_6$ are not obvious. This is strongly different from the behavior appearing in $Bi_2Sn_2O_7$ [31].

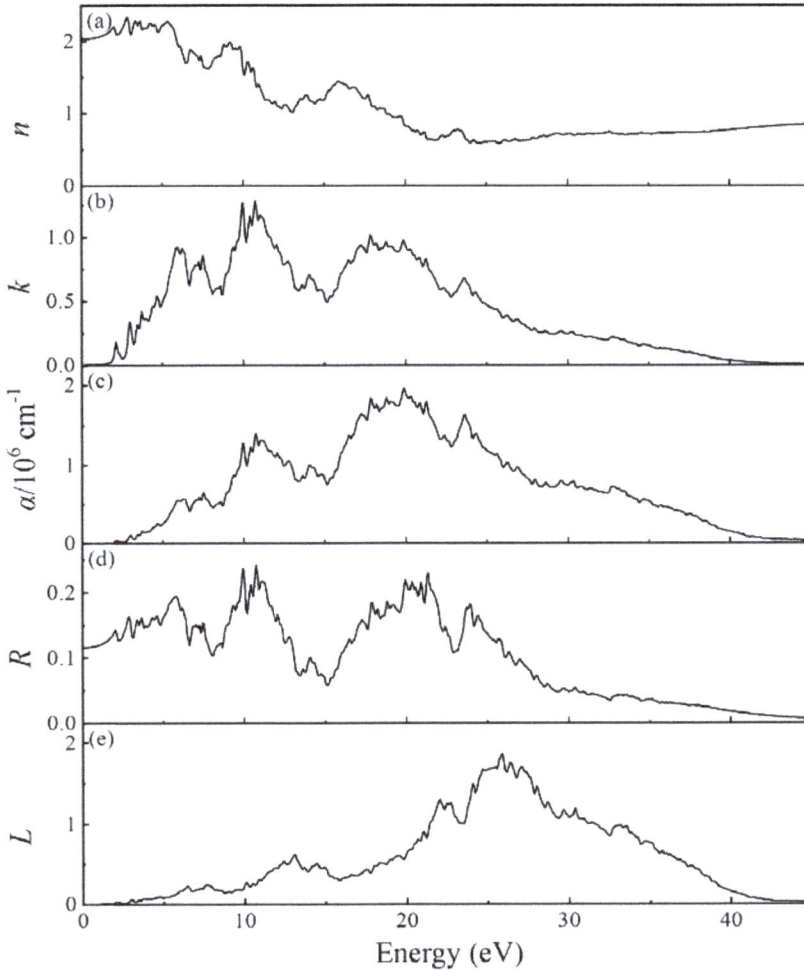

Figure 6. Calculated optical properties of $MgBi_2O_6$: (**a**) complex refractive index *n*, (**b**) extinction coefficient *k*, (**c**) absorption coefficient α, (**d**) reflectivity *R*, and (**e**) electron energy loss function *L*.

3.4. Mechanical Properties of $MgBi_2O_6$

To investigate the mechanical properties of tetragonal $MgBi_2O_6$, the six independent elastic constants C_{ij} are calculated by applying finite distortions of the lattice. The results are listed in Table 3. For the tetragonal system, the mechanical stability criterion [32] is given by the following relationships:

$$(C_{11} - C_{12}) > 0, \ (C_{11} + C_{33} - 2C_{13}) > 0,$$

$$C_{11} > 0, \ C_{33} > 0, \ C_{44} > 0, \ C_{66} > 0,$$

$$(2C_{11} + C_{33} + 2C_{12} + 4C_{13}) > 0, \tag{2}$$

The calculated values of C_{ij} in Table 3 satisfy the stability criterion mentioned above, indicating that the tetragonal $MgBi_2O_6$ has mechanical stability. As listed in Table 3, C_{11} is significantly smaller than C_{33}, indicating that the chemical bonding strength in the (100) and (010) directions is significantly weaker than the bonding strength in the (001) direction. In addition, C_{44} is obviously smaller than C_{66}, which demonstrates that it is easier for shear deformation to occur along the (001) direction in comparison with the (010) direction. The shear elastic anisotropy of the material can be estimated by the relation $A = 2C_{66}/(C_{11} - C_{12})$. Typically, if A has a value of 1, meaning that the material is isotropic. The more the value of A deviates from 1, the elastic anisotropy would be more prominent. For $MgBi_2O_6$, the calculated A value is 4.6, indicating that $MgBi_2O_6$ is highly anisotropic.

Table 3. Calculated elastic constants C_{ij}, bulk modulus B, shear modulus G, Young's modulus E, and Poisson's ratio v of $MgBi_2O_6$.

Elastic Constants (GPa)						Mechanical Moduli (GPa)			v
C_{11}	C_{12}	C_{13}	C_{33}	C_{44}	C_{66}	B	G	E	
171.8	110.4	98.1	280.3	66.2	141.9	137.7	76.2	193.0	0.27

For the tetragonal system, the bulk modulus B and the shear modulus G are calculated as follows:

$$B = (2C_{11} + C_{33} + 2C_{12} + 4C_{13})/9, \tag{3}$$

$$G = (2C_{11} + C_{33} - C_{12} - 2C_{13} + 6C_{44} + 3C_{66})/15, \tag{4}$$

Young's modulus E and Poisson's ratio v can be estimated from the bulk and shear moduli

$$E = 9BG/(G + 3B), \tag{5}$$

$$v = (3B - 2G)/[2(3B + G)], \tag{6}$$

The calculated B, G, E, and v are listed in Table 3. The B/G value is about 1.81. According to Pugh's criteria of brittleness and ductility [33], $MgBi_2O_6$ exhibits some toughness, but it is not obvious, which is consistent with the Poisson's ratio $v = 0.27$. In addition, the elastic anisotropy of $MgBi_2O_6$ can be directly determined by the direction-dependent Young's modulus. The Young's modulus in a specific direction can be expressed by the elastic compliances (s'):

$$E' = \frac{1}{s'}, \tag{7}$$

with

$$s' = A_{1i}A_{1j}A_{1k}A_{1l}s_{ijkl}, \tag{8}$$

where A is the matrix associated with the change of axes:

$$A_{11} = \cos\theta \sin\varphi, \ A_{12} = \sin\theta \sin\varphi, \ A_{13} = \cos\varphi, \tag{9}$$

Using the reduction of s_{ijkl} for the tetragonal crystal class [34], the reduced Young's modulus with orientation can be expressed as follows:

$$E' = \frac{1}{\left\{s_{11} + [s_{66} - 2(s_{11} - s_{12})]\frac{\sin^2\theta}{4}\right\}\sin^4\theta + s_{33}\cos^4\theta + (2s_{13} + s_{44})\frac{\sin^2 2\varphi}{4}}, \tag{10}$$

The calculated directional dependence of Young's modulus depicted in Figure 7 is a significantly distorted spherical shape. The calculated values along the [001], [110], and [-110] directions are obviously larger than those along the [100] and [010] directions. Tetragonal $MgBi_2O_6$ exhibits the highly elastic anisotropy.

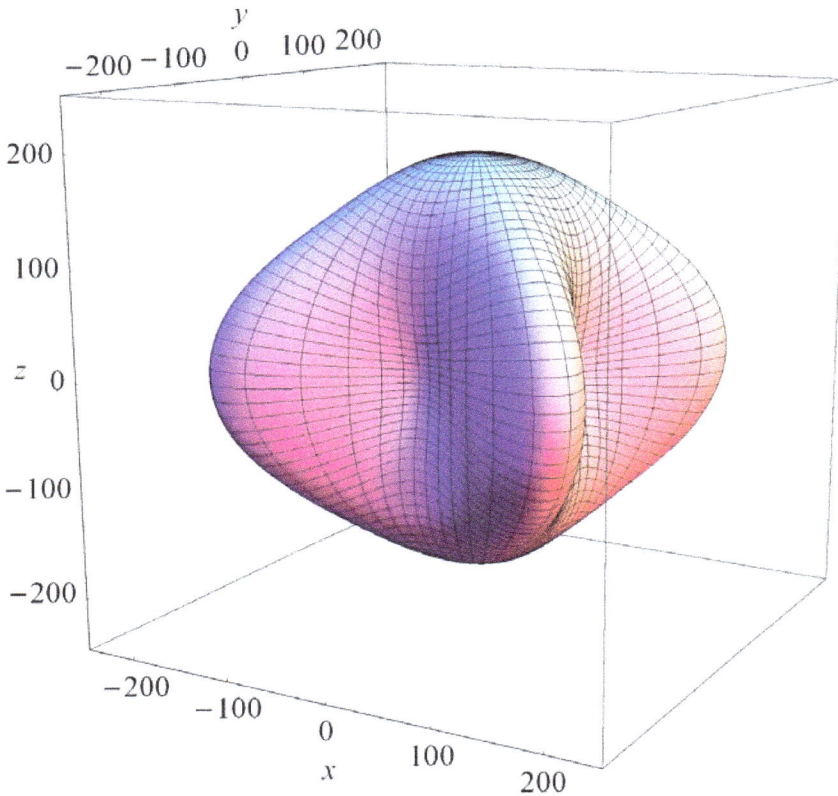

Figure 7. Directional dependence of Young's modulus (in GPa) of $MgBi_2O_6$.

3.5. Lattice Dynamical Properties of $MgBi_2O_6$

The phonon dispersion curves along high symmetry directions and the phonon density of states are shown in Figure 8. There are 18 atoms in $MgBi_2O_6$ cells, so there is a total of 54 vibration modes in the phonon spectrum, including three acoustic modes and 51 optical modes. The calculated phonon spectrum shows no imaginary frequency, indicating the dynamical stability of $MgBi_2O_6$. Along *F-Q* and *Q-Z* paths in the Brillouin zone, the vibration modes are double degenerate. The acoustic modes phonons reflect the vibration of the centroid of the original cell and occupy the 0–3 THz frequency region. Among the three acoustic modes passing through the Γ point, the longitudinal mode has higher frequency in comparison with the other two transverse acoustic modes. There is no gap at the frequency range of about 3 THz between the longitudinal acoustic and transverse optical modes. Thus, phonons can transition from the acoustic mode to the optical mode without any momentum transfer [35]. Further combining with the calculated PDOS shown in Figure 8, we can find that the coupled vibrations of Bi, Mg, and O atoms including the O-Bi-O and O-Mg-O bending vibrations in BiO_6 and MgO_6 octahedrons within the frequency range from 3 to 9.8 THz and the Bi-O and Mg-O stretching vibrations in the frequency range from 12 to 16.4 THz are obvious. The phonon branches above 16.4 THz are completely ascribed to the vibrations of O atoms.

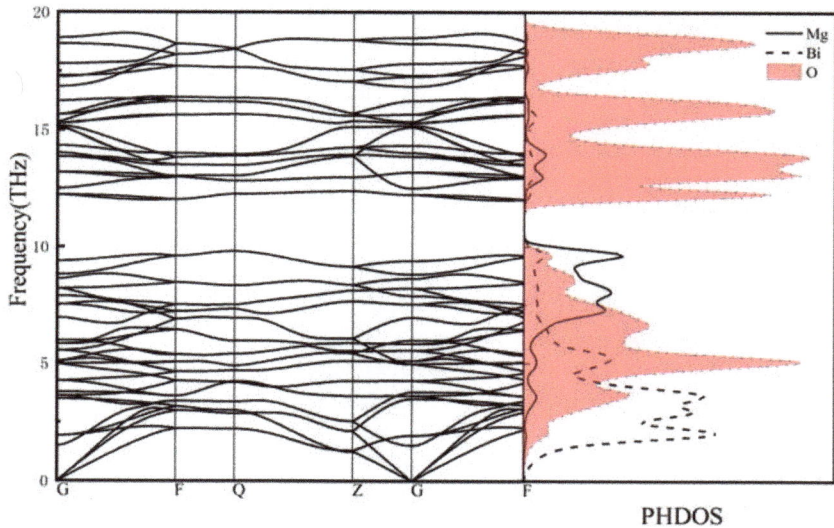

Figure 8. Calculated phonon dispersion curves and phonon density of states of $MgBi_2O_6$.

4. Conclusions

In this work, we have investigated the structural, electronic, optical, mechanical, and lattice dynamical properties of the trirutile-type $MgBi_2O_6$ in detail using the first-principles calculations. The calculated band gap of $MgBi_2O_6$ from the Heyd-Scuseria-Ernzerhof hybrid functional PBE0 electrical is about 1.62 eV and consistent with the experimental data (1.6~1.8 eV). The calculated effective masses show that the mobility of holes at conduction band is obviously slower than that of electrons at valence band, indicating high separation efficiency of electrons and holes in $MgBi_2O_6$. The calculated results of optical parameters clearly show that $MgBi_2O_6$ has strong light response in the visible light region and can be used as a light absorbing material. The calculated elastic constants and phonon dispersion clearly show that $MgBi_2O_6$ is mechanically and dynamically stable. Moreover, $MgBi_2O_6$ exhibits significantly elastic anisotropy.

Author Contributions: Conceptualization, C.H. and Y.Z.; methodology, D.W.; software, D.W.; validation, L.L., D.W. and Y.Z.; formal analysis, L.L.; investigation, L.L.; resources, C.H.; data curation, L.L.; writing—original draft preparation, L.L.; writing—review and editing, D.W., Y.Z., C.H.; visualization, L.L.; supervision, C.H.; project administration, C.H.; funding acquisition, C.H.

Funding: This research was funded by the National Natural Science Foundation of China, grant number 11464008, the Natural Science Foundation of Guangxi Zhuang Autonomous Region, Grant Number 2014GXNSFGA118001 and 2016GXNSFGA380001, the Talents Project of Guilin University of Electronic Technology, and Guangxi Key Laboratory of Information Materials, Grant Number 1210908-215-Z and 171034-Z.

Conflicts of Interest: The authors declare no conflict of interest.

References

1. Wang, J.; Neaton, J.; Zheng, H.; Nagarajan, V.; Ogale, S.; Liu, B.; Viehland, D.; Vaithyanathan, V.; Schlom, D.; Waghmare, U. Epitaxial BiFeO3 multiferroic thin film heterostructures. *Science* **2003**, *299*, 1719. [CrossRef] [PubMed]
2. Jeen, H.; Singh-Bhalla, G.; Mickel, P.R.; Voigt, K.; Morien, C.; Tongay, S.; Hebard, A.; Biswas, A. Growth and characterization of multiferroic BiMnO3 thin films. *J. Appl. Phys.* **2011**, *109*, 074104. [CrossRef]

3. Subramanian, M.A.; Torardi, C.C.; Calabrese, J.C.; Gopalakrishnan, J.; Morrissey, K.J.; Askew, T.R.; Flippen, R.B.; Chowdhry, U.; Sleight, A.W. A new high-temperature superconductor: $Bi_2Sr_{3-x}Ca_xCu_2O_{8+y}$. *Science* **1988**, *239*, 1015. [CrossRef] [PubMed]

4. Tallon, J.L.; Buckley, R.G.; Gilberd, P.W.; Presland, M.R.; Brown, I.W.M.; Bowden, M.E.; Christian, L.A.; Goguel, R. High-Tc superconducting phases in the series $Bi_{2.1}(Ca, Sr)_{n+1}Cu_nO_{2n+4+\delta}$. *Nature* **1988**, *333*, 153. [CrossRef]

5. Shaddad, M.N.; Cardenas-Morcoso, D.; Arunachalam, P.; García-Tecedor, M.; Ghanem, M.A.; Bisquert, J.; Al-Mayouf, A.; Gimenez, S. Enhancing the optical absorption and interfacial properties of $BiVO_4$ with Ag_3PO_4 nanoparticles for efficient water splitting. *J. Phy. Chem. C* **2018**, *122*, 11608. [CrossRef]

6. Meng, X.; Zhang, Z. Bismuth-based photocatalytic semiconductors: Introduction, challenges and possible approaches. *J. Mol. Catal. A Chem.* **2016**, *423*, 533. [CrossRef]

7. Xu, D.; Yang, H.; Zhang, X.; Zhang, S.; He, R. Bi_2O_3 cocatalyst improving photocatalytic hydrogen evolution performance of TiO_2. *Appl. Surf. Sci.* **2017**, *400*, 530. [CrossRef]

8. Kudo, A.; Omori, K.; Kato, H. A novel aqueous process for preparation of crystal form-controlled and highly crystalline $BiVO_4$ powder from layered vanadates at room temperature and its photocatalytic and photophysical properties. *J. Am. Chem. Soc.* **1999**, *121*, 11459. [CrossRef]

9. Fu, H.; Pan, C.; Yao, W.; Zhu, Y. Visible-light-induced degradation of rhodamine B by nanosized Bi_2WO_6. *J. Phys. Chem. B* **2005**, *109*, 22432. [CrossRef]

10. Wu, J.; Huang, F.; Lü, X.; Chen, P.; Wan, D.; Xu, F. Improved visible-light photocatalysis of nano-$Bi_2Sn_2O_7$ with dispersed s-bands. *J. Mater. Chem.* **2011**, *21*, 3872. [CrossRef]

11. Shang, S.L.; Sheng, G.; Wang, Y.; Chen, L.Q.; Liu, Z.K. Elastic properties of cubic and rhombohedral $BiFeO_3$ from first-principles calculations. *Phys. Rev. B* **2009**, *80*, 052102. [CrossRef]

12. Zhai, L.J.; Wang, H.Y. The magnetic and multiferroic properties in $BiMnO_3$. *J. Magn. Magn. Mater.* **2017**, *426*, 188. [CrossRef]

13. Kako, T.; Zou, Z.; Katagiri, M.; Ye, J. Decomposition of organic compounds over $NaBiO_3$ under visible light irradiation. *Chem. Mater.* **2007**, *19*, 198. [CrossRef]

14. Gong, J.; Lee, C.S.; Kim, E.J.; Kim, J.H.; Lee, W.; Chang, Y.S. Self-generation of reactive oxygen species on crystalline $AgBiO_3$ for the oxidative remediation of organic pollutants. *ACS Appl. Mater. Interfaces* **2017**, *9*, 28426. [CrossRef] [PubMed]

15. Chouhan, A.S.; Athresh, E.; Ranjan, R.; Raghavan, S.; Avasthi, S. $BaBiO_3$: A potential absorber for all-oxide photovoltaics. *Mater. Lett.* **2018**, *210*, 218. [CrossRef]

16. Tang, J.; Zou, Z.; Ye, J. Efficient photocatalysis on $BaBiO_3$ driven by visible light. *J. Phys. Chem. C* **2007**, *111*, 12779. [CrossRef]

17. Kumada, N.; Takahashi, N.; Kinomura, N.; Sleight, A. Preparation of ABi_2O_6(A= Mg, Zn) with the trirutile-type structure. *Mater. Res. Bull.* **1997**, *32*, 1003. [CrossRef]

18. Mizoguchi, H.; Bhuvanesh, N.S.P.; Woodward, P.M. Optical and electrical properties of the wide gap, n-type semiconductors: $ZnBi_2O_6$ and $MgBi_2O_6$. *Chem. Commun.* **2003**, 1084. [CrossRef]

19. Takei, T.; Haramoto, R.; Dong, Q.; Kumada, N.; Yonesaki, Y.; Kinomura, N.; Mano, T.; Nishimoto, S.; Kameshima, Y.; Miyake, M. Photocatalytic activities of various pentavalent bismuthates under visible light irradiation. *J. Solid State Chem.* **2011**, *184*, 2017. [CrossRef]

20. Zhong, L.; Hu, C.; Zhuang, J.; Zhong, Y.; Wang, D.; Zhou, H. $AgBr/MgBi_2O_6$ heterostructured composites with highly efficient visible-light-driven photocatalytic activity. *J. Phys. Chem. Solids* **2018**, *117*, 94. [CrossRef]

21. Zhang, C.; Kou, L.; He, T.; Jiao, Y.; Liao, T.; Bottle, S.; Du, A. First principles study of trirutile magnesium bismuth oxide: Ideal bandgap for photovoltaics, strain-mediated band-inversion and semiconductor-to-semimetal transition. *Comput. Mater. Sci.* **2018**, *149*, 158. [CrossRef]

22. Kresse, G.; Furthmüller, J. Efficient iterative schemes for ab initio total-energy calculations using a plane-wave basis set. *Phys. Rev. B* **1996**, *54*, 11169. [CrossRef]

23. Kresse, G.; Furthmüller, J. Efficiency of ab-initio total energy calculations for metals and semiconductors using a plane-wave basis set. *Comput. Mater. Sci.* **1996**, *6*, 15. [CrossRef]

24. Blöchl, P.E. Projector augmented-wave method. *Phys. Rev. B* **1994**, *50*, 17953. [CrossRef]

25. Kresse, G.; Joubert, D. From ultrasoft pseudopotentials to the projector augmented-wave method. *Phys. Rev. B* **1999**, *59*, 1758. [CrossRef]

Appl. Sci. **2019**, *9*, 1267

26. Perdew, J.P.; Burke, K.; Ernzerhof, M. Generalized gradient approximation made simple. *Phys. Rev. Lett.* **1996**, *77*, 3865. [CrossRef] [PubMed]

27. Paier, J.; Hirschl, R.; Marsman, M.; Kresse, G. The Perdew–Burke–Ernzerhof exchange-correlation functional applied to the G2-1 test set using a plane-wave basis set. *J. Chem. Phys.* **2005**, *122*, 234102. [CrossRef]

28. Paier, J.; Marsman, M.; Hummer, K.; Kresse, G.; Gerber, I.C.; Ángyán, J.G. Screened hybrid density functionals applied to solids. *J. Chem. Phys.* **2006**, *124*, 154709. [CrossRef]

29. Togo, A.; Tanaka, I. First principles phonon calculations in materials science. *Scr. Mater.* **2015**, *108*, 1. [CrossRef]

30. Zhang, H.J.; Liu, L.; Zhou, Z. Towards better photocatalysts: First-principles studies of the alloying effects on the photocatalytic activities of bismuth oxyhalides under visible light. *Phys. Chem. Chem. Phys.* **2012**, *14*, 1286. [CrossRef] [PubMed]

31. Hu, C.H.; Yin, X.H.; Wang, D.H.; Zhong, Y.; Zhou, H.Y.; Rao, G.H. First-principles studies of electronic, optical, and mechanical properties of γ-$Bi_2Sn_2O_7$. *Chin. Phys. B* **2016**, *25*, 067801. [CrossRef]

32. Wu, Z.J.; Zhao, E.J.; Xiang, H.P.; Hao, X.F.; Liu, X.J.; Meng, J. Crystal structures and elastic properties of superhard IrN_2 and IrN_3 from first principles. *Phys. Rev. B* **2007**, *76*, 054115. [CrossRef]

33. Pugh, S.F. XCII. Relations between the elastic moduli and the plastic properties of polycrystalline pure metals. *Philos. Mag.* **1954**, *45*, 823. [CrossRef]

34. Authier, A. *International Tables for Crystallography: Vol. D, Physical Properties of Crystals*; Kluwer Academic Publishers: Dordrecht, The Netherland, 2003; pp. 83–84.

35. Shrivastava, D.; Sanyal, S.P. Structural phase transition, electronic and lattice dynamical properties of half-Heusler compound CaAuBi. *J. Alloys Compd.* **2018**, *745*, 240. [CrossRef]

applied
sciences

MDPI

Article

Theoretical Investigations on the Mechanical, Magneto-Electronic Properties and Half-Metallic Characteristics of ZrRhTiZ (Z = Al, Ga) Quaternary Heusler Compounds

Wenbin Liu [1], Xiaoming Zhang [1], Hongying Jia [2], Rabah Khenata [3], Xuefang Dai [1,*] and Guodong Liu [1,*]

[1]   School of Material Science and Engineering, Hebei University of Technology, Tianjin 300130, China; wbliu1106@126.com (W.L.); zhangxiaoming87@hebut.edu.cn (X.Z.)
[2]   Peter Grünberg Institut and Institute for Advanced Simulation, Forschungszentrum Jülich, and JARA, 52425 Jülich, Germany; h.jia@fz-juelich.de
[3]   Laboratoire de Physique Quantique et de Modelisation Mathematique (LPQ3M), Departement de Technologie, Universite de Mascara, Mascara 29000, Algeria; khenata_rabah@yahoo.fr
*   Correspondence: xuefangdai@126.com (X.D.); gdliu1978@126.com (G.L.)

Received: 3 January 2019; Accepted: 29 January 2019; Published: 1 March 2019

Abstract: The electronic, magnetic, and mechanical properties were investigated for ZrRhTiZ (Z = Al, Ga) quaternary Heusler compounds by employing first-principles calculations framed fundamentally within density functional theory (DFT). The obtained electronic structures revealed that both compounds have half-metallic characteristics by showing 100% spin polarization near the Fermi level. The half-metallicity is robust to the tetragonal distortion and uniform strain of the lattice. The total magnetic moment is 2 μ_B per formula unit and obeys the Slater-Pauling rule, $M_t = Z_t - 18$ (M_t and Z_t represent for the total magnetic moment and the number of total valence electrons in per unit cell, respectively). The elastic constants, formation energy, and cohesive energy were also theoretically calculated to help understand the possibility of experimental synthesis and the mechanical properties of these two compounds.

Keywords: half-metallic materials; first-principles calculations; quaternary Heusler compound

1. Introduction

Half-metallic materials (HMMs) [1] can provide completely spin-polarized conducting electrons due to their unique electronic band structure which shows metallic characteristics in one spin channel and semiconducting/insulating properties in the other spin channel. Hence, HMMs are very much appreciated as key materials for providing high spin polarized carriers for spintronics devices.

As one of the important HMM families, Heusler compounds have a special importance due to their high Curie temperature, tunable electronic structure, and wide change of lattice constant [2,3]. Moreover, from the previous reports concerning pseudo-quaternary or ternary Heusler compounds with half-metallicity, it was found that the materials containing 4d or 5d transition metal elements generally demonstrate a wide band gap implying a robust half-metallicity in a quite uniform strain or tetragonal distortion, which is quite suitable for practical applications. Therefore, in recent years, the investigations on the Heusler compounds containing 4d or 5d transition metal elements have become the research focus [4–9]. Although many ternary Heusler compounds containing 4d transition metal elements have been reported in the literature, the quaternary Heusler compounds containing 4d or 5d transition metal elements, especially, the stoichiometric quaternary Heusler (SQH) compounds with half-metallicity have been rarely reported. Hence, it is important to further investigate and search for new half-metallic SQH materials containing 4d or 5d transition metal elements.

In this work, we investigate two new SQH compounds, ZrRhTiAl and ZrRhTiGa, for the first time. Our prime aim in this work is to investigate the electronic and magnetic properties of both compounds and their half-metallic (HM) stability under uniform strain and tetragonal distortion. The elastic constants, formation energy, and cohesive energy were also presented to help understand the possibility of experimental synthesis and the mechanical properties of these two compounds.

2. Method of Calculations

In this work, the calculations were performed by employing CASTEP computational code [10] framed fundamentally within DFT [11]. In all the calculations, the ultra-soft pseudo-potential approach with plane wave basis set was used, and the generalized-gradient-approximation (GGA) was adopted for the exchange–correction energy functional part of the total energy [12,13]. For the expansion of electronic wave functions, the plane wave basis set expansion approach was used. To truncate the plane wave basis set expansion to attain the required convergence criterion, an energy cut-off of 450 eV and a mesh of $12 \times 12 \times 12$ k-points for the Brillouin zone sampling was used. Within our above said applied parameters, calculations ensured a high-level total energy convergence with less than a tolerance of 1×10^{-6} eV per atom for both ZrRhTiZ (Z = Al, Ga) compounds.

3. Results and Discussions

The quaternary Heusler compounds have a LiMgPdSb-type structure, which belongs to the F$\overline{4}$3m (No.216) space group. There are four crystallographic sites in Heusler compounds. For quaternary Heusler compounds, there are three possible atomic arrangements depending on each crystallographic site occupation. In this present study, the atomic arrangement properties of ZrRhTiZ (Z = Al, Ga) compounds were investigated by calculating the total energy in ferromagnetic (FM) and non-ferromagnetic (NM) states. The calculated results show that for the ZrRhTiZ (Z = Al, Ga) compound, the most stable case is that Zr and Ti atoms with less valence electrons enter into the Wyckoff sites 4a(0, 0, 0) and 4c(0.25, 0.25, 0.25), Rh atoms with more valence electrons occupy the 4b(0.5, 0.5, 0.5) site, and the main group of atoms, Z atoms, tend to locate at the 4d(0.75, 0.75, 0.75) site. The corresponding simulated crystal structure is shown in Figure 1. The stable arrangement of ZrRhTiZ (Z = Al, Ga) Heusler compounds is like the combination of inverse Zr$_2$RhZ [14] and Ti$_2$RhZ (Z = Al, Ga) [15] Heusler compounds and found to be similar to the other Zr-based SQH compounds [16,17].

Figure 2 shows the dependence of the total energy on the lattice parameters in FM and NM states for ZrRhTiZ (Z = Al, Ga) compounds. One can see that both compounds are more stable in FM than NM states according to the viewpoint of the total energy minimization. The equilibrium lattice constants corresponding to the ground state energy are 6.47 Å and 6.45 Å for ZrRhTiAl and ZrRhTiGa, respectively.

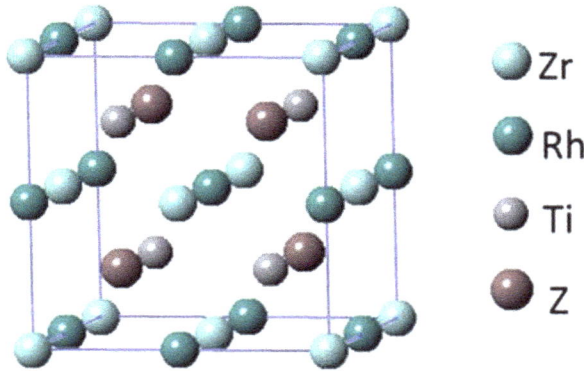

Figure 1. Crystal structure of ZrRhTiZ (Z = Al, Ga) quaternary Heusler compounds.

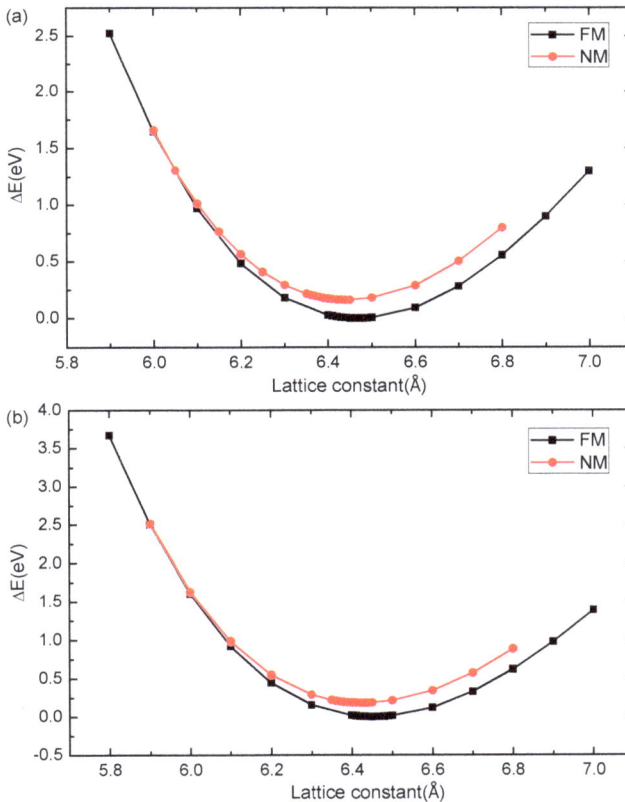

Figure 2. Total energy as a function of the lattice constant in the non-ferromagnetic (NM) and ferromagnetic (FM) states for ZrRhTiAl (**a**) and ZrRhTiGa (**b**) compounds. The lowest energy in the FM state is set as zero point.

The spin-projected band structures in the reduced Brillouin zone along the high symmetry directions were calculated under the equilibrium lattice constant and plotted in Figure 3 for ZrRhTiAl and ZrRhTiGa compounds. From Figure 3, an indirect band gap can be clearly observed in the spin-down channel, and the Fermi level is located within the band gap. In the spin-up channel, the valence and conduction bands are overlapping, and the Fermi level is intersecting them for ZrRhTiZ (Z = Al, Ga) compounds, which indicates that both compounds are half-metallic materials. The band gap (E_{bg}) and HM band gap (E_{HM}) in the spin-down channel are listed in Table 1. The HM band gap is the minimum energy required to flip a minority of spin electrons from the valence band maximum (VBM) to the majority spin Fermi level. The E_{bg} is 0.432 eV for ZrRhTiAl and 0.541 eV for ZrRhTiGa. Usually, the large E_{bg} can be considered as evidence of the robustness of half-metallicity to lattice distortion. The Mulliken atomic populations quantify the charge transfer from one atom to another one and are listed in Table 2 for ZrRhTiZ (Z = Al, Ga) compounds. It is clear that the charge transfer from Ti to Rh is 0.69e and Zr to Al is 0.08e in ZrRhTiAl, and the charge transfer from Ti to Zr is 0.27e and Ga to Rh is 0.93e in ZrRhTiGa.

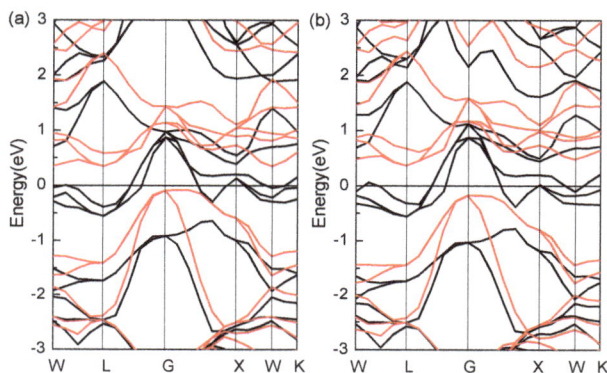

Figure 3. The calculated band structures for ZrRhTiAl (**a**) and ZrRhTiGa (**b**). (The black lines represent the spin-up channel, and the red lines the spin-down channel.).

Table 1. The equilibrium lattice constants (Å), total magnetic moments (μ_B), atomic magnetic moments (μ_B), valence band maximum (eV), conduction band minimum (eV), band gap (eV), half-metallic (HM) band gap (eV), and the number of valence electrons for ZrRhTiZ (Z = Al, Ga) compounds.

Compound	Total	Zr	Rh	Ti	Z	a (Å)	CBM	VBM	E_{bg}	E_{HM}	Z_t	S-P rule	P (%)
ZrRhTiAl	2.00	1.26	−0.32	1.18	−0.14	6.47	0.350	−0.082	0.432	0.082	20	$M_t = Z_t - 18$	100
ZrRhTiGa		1.22	−0.32	1.32	−0.22	6.45	0.360	−0.181	0.541	0.181			

Table 2. Mulliken population analysis of ZrRhTiAl and ZrRhTiGa.

Species	Atom	s	p	d	Total	Charge(e)
ZrRhTiAl	Zr	2.54	6.37	3.02	11.93	0.07
	Rh	0.89	0.60	8.20	9.69	−0.69
	Ti	2.53	6.12	2.65	11.30	0.70
	Al	1.05	2.03	0.00	3.08	−0.08
ZrRhTiGa	Zr	2.64	6.67	2.96	12.27	−0.27
	Rh	0.97	0.76	8.20	9.93	−0.93
	Ti	2.70	6.37	2.65	11.73	0.27
	Ga	−0.17	2.26	9.99	12.08	0.92

Figure 4 shows the total density of state (TDOS) and partial density of state (PDOS) patterns. It is clear that the main contributions to the TDOS near the Fermi level are the strong d-d hybridization among the Zr-4d state, Ti-3d state, and Rh-4d state. The spin polarization (P) at Fermi energy is

100% for ZrRhTiZ (Z = Al, Ga), for the Fermi level lies in a band gap in the spin-down channel. For Heusler compounds with half-metallicity, the origin of the half-metallic band gap is usually owed to the hybridization of the d electrons between the transition metal atoms. Through the classical molecular orbital approach, Zr-4d and Rh-4d hybridization within tetrahedral symmetry has been considered, as presented in Figure 5. In ZrRhTiZ (Z = Al, Ga) Heusler compounds, the 4d orbitals hybridization of Rh and Zr atoms generated five bonding bands ($3t_{2g}$ and $2e_g$) and five non-bonding bands ($2e_u$ and $3t_{1u}$). Then, the 3d orbitals of Ti atom hybridize with the five bonding 4d hybridized orbitals ($3t_{2g}$ and $2e_g$) of Rh and Zr atom, while the five non-bonding hybridized 4d orbitals ($2e_u$ and $3t_{1u}$) still hold with no hybridization. Finally, the distribution of the 15 d orbitals in the spin-down channel can be determined, i.e., $3 \times t_{2g}$, $2 \times e_g$, $2 \times e_u$, $3 \times t_{1u}$, $3 \times t_{2g}$, and $2 \times e_g$ are arranged from high energy to low energy. The main-group of Z atoms generate $1 \times s$ orbital and $3 \times p$ orbitals which are distributed below the 15 hybridized d orbitals and totally occupied in ZrRhTiZ (Z = Al, Ga) compounds. The triple-degeneracy t_{1u} states are not occupied in the spin-down direction, and a band gap is formed between t_{1u} (non-bonding) and t_{2g} (bonding) orbitals in the spin-down channel of ZrRhTiZ (Z = Al, Ga) compounds. Similarly, from Table 1, it can be seen that the number of valence electrons is 20 per ZrRhTiZ (Z = Al, Ga) unit cell, and the calculated total magnetic moment is 2 μ_B, which follows the Slater-Pauling rule $M_t = Z_t - 18$ and is also consistent with the reported Zr-based SQH materials [4–7,16–18].

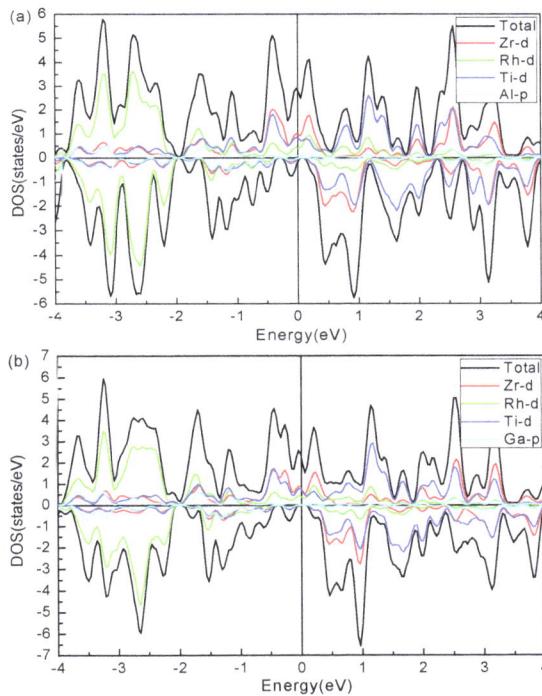

Figure 4. Calculated total density of states and partial density of states of ZrRhTiAl (**a**) and ZrRhTiGa (**b**).

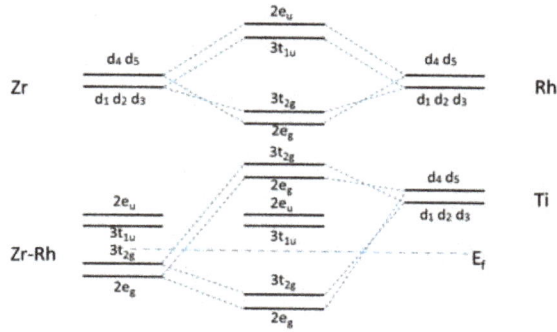

Figure 5. Schematic diagram of possible d-d hybridization between the transition-metal elements Zr-4d, Rh-4d, and Ti-3d in the ZrRhTiZ (Z = Al, Ga) compounds. The main group of elements Z (Z = Al, Ga) are not taken into account because the sp-bands are located at deep energy levels and barely contribute to the gap formation.

Figure 6 displays total magnetic moment and atomic magnetic moments as a function of lattice constant for ZrRhTiAl and ZrRhTiGa compounds. It can be noted that the total magnetic moment is constant at $2\mu_B$ over the range of the lattice constant, from 6.09 (6.10) to 6.61 (6.75) Å for ZrRhTiAl (ZrRhTiGa) compounds. The atomic magnetic moment is appreciably increasing with the lattice constant values for the Rh and Ti atoms, whereas the atomic magnetic moment of Zr as well as Al/Ga was not found to change appreciably over the wide range of lattice constants. The constant total magnetic moment can be attributed to the antiparallel arrangement of the magnetic moments of Rh and Ti atoms. The main contributors to the magnetic moment are Zr and Ti atoms. The Rh and Z atoms only carry small magnetic moments antiparallelly aligned to those of the Zr and Ti atoms.

The half-metallicity is very much sensitive to the change of lattice constant. Therefore, it is very useful to acquire knowledge about the sensitivity of the half-metallicity to the uniform strain (US) for the ZrRhTiZ (Z = Al, Ga) compounds. Here, the CBM (conduction band minimum) and VBM in the spin-down channel at different lattice constants have been recorded to show the effects of the US on half-metallic behavior, as plotted in Figure 7. Obviously, the half-metallicity can be maintained in the lattice constant range of 6.09–6.61 Å for ZrRhTiAl and 6.10–6.75 Å for ZrRhTiGa. In addition, the effects of the US on the E_{bg} and E_{HM} are also shown in Figures 8 and 9 for the ZrRhTiZ (Z = Al, Ga) compounds, respectively. The obtained results revealed both E_{HM} and E_{bg} first increase and then decrease with the increasing lattice constant.

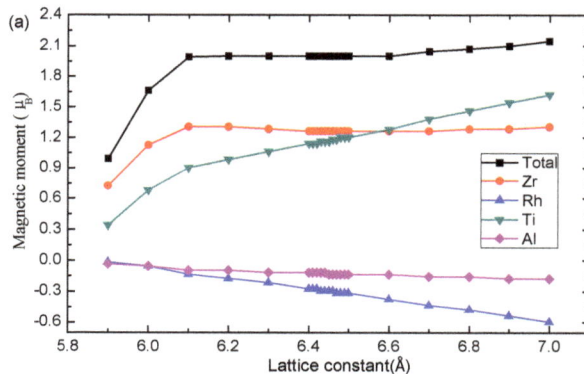

Figure 6. *Cont.*

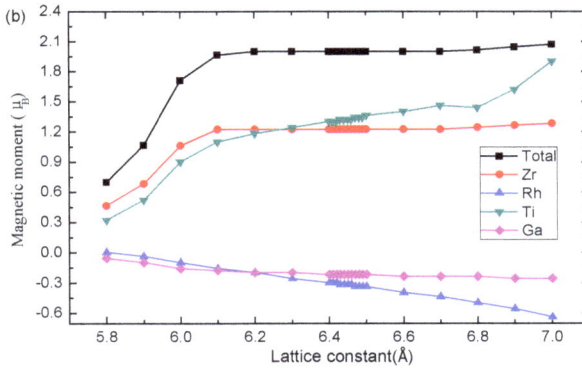

Figure 6. Plots for total spin magnetic moment and atomic spin magnetic moment as a function of lattice constant for ZrRhTiAl (**a**) and ZrRhTiGa (**b**) compounds.

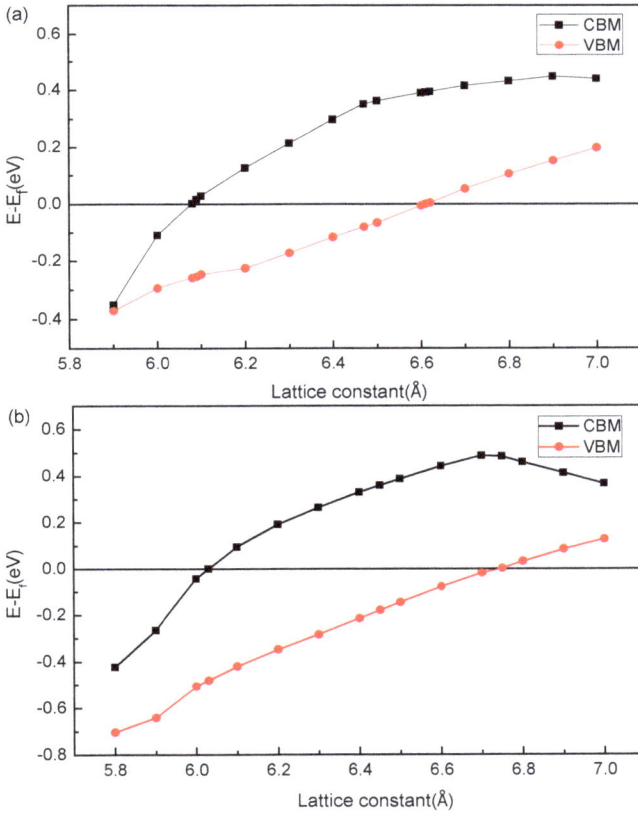

Figure 7. Conduction band minimum (CBM) and valence band maximum (VBM) in the spin-down channel as a function of lattice constant (US) for ZrRhTiAl (**a**) and ZrRhTiGa (**b**).

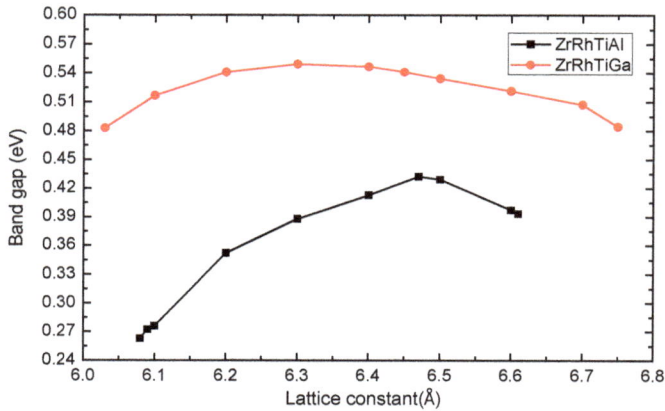

Figure 8. Band gaps as a function of the lattice constant (US) for ZrRhTiAl and ZrRhTiGa.

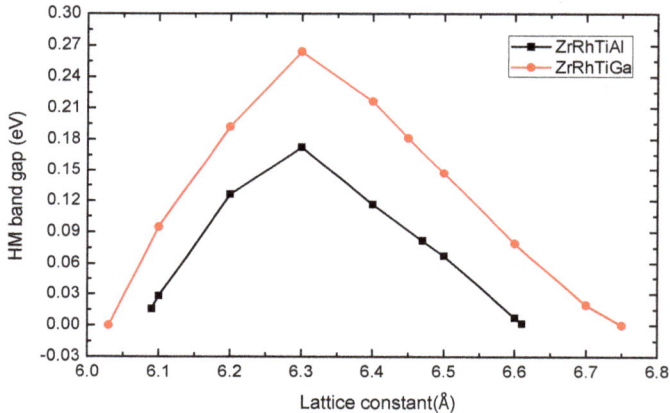

Figure 9. The half-metallic band-gap as a function of the lattice constant (US) for ZrRhTiAl and ZrRhTiGa.

The half-metallicity of most Heusler compounds will be influenced and even be broken by the tetragonal distortion (TD) of the lattice. To analyze the effect of TD on half-metallicity and magnetic moments, the unit-cell volume was fixed at the equilibrium bulk volume and then the c/a ratio changed. In Figures 10 and 11, the TD effects on the half-metallicity properties and magnetic moments are described by the magnetic moment and the CBM and VBM dependence on c/a ratio. It is clear that the total and atomic magnetic moments of ZrRhTiZ (Z = Al, Ga) are nearly unchanged, and the half-metallicity can be kept in the c/a ratio range of 0.96–1.06 for ZrRhTiAl and 0.92–1.12 for ZrRhTiGa, respectively. We further plotted the curves of E_{bg} and E_{HM} as a function of c/a ratio, as shown in Figures 12 and 13, respectively. The E_{bg} and E_{HM} have a similar variation tendency and showed a rise-fall characteristic with increasing c/a ratio. The largest E_{bg} and E_{HM} occur in the cubic structure (c/a = 1) instead of tetragonal phase.

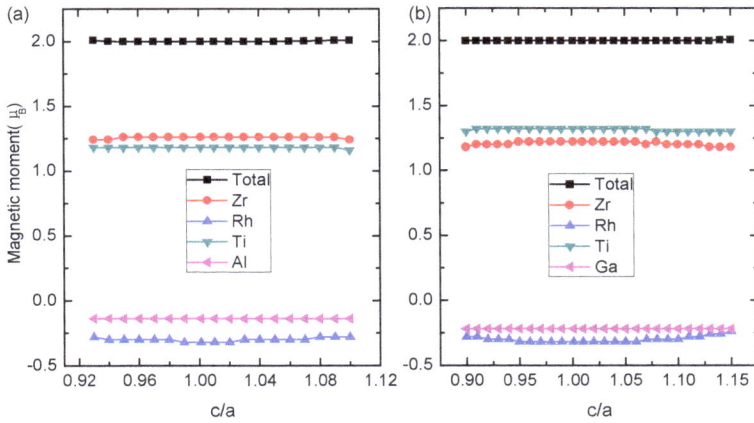

Figure 10. Total spin magnetic moment and atomic spin magnetic moment as a function of c/a ratio (TD) for ZrRhTiAl (**a**) and ZrRhTiGa (**b**).

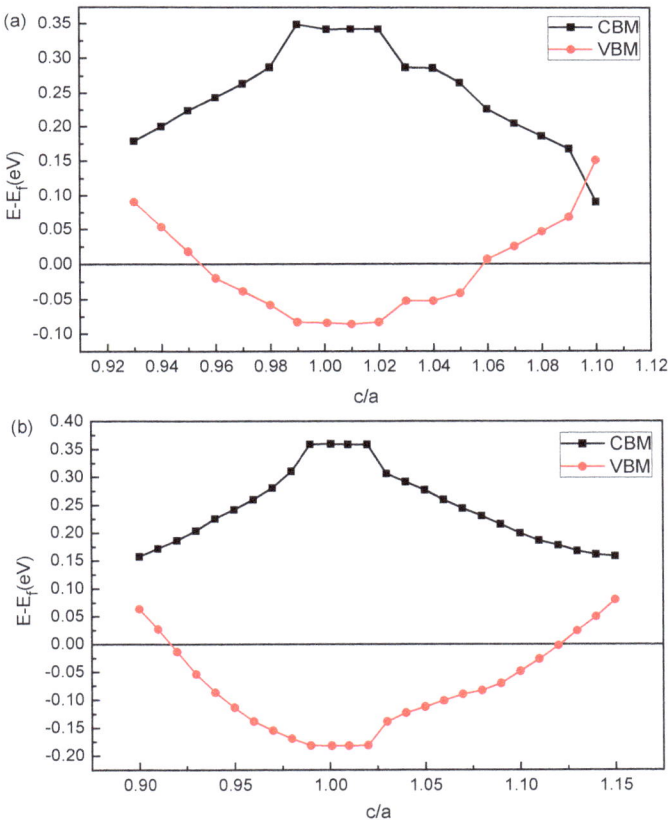

Figure 11. CBM and VBM in the minority channel as a function of c/a ratio (TD) for ZrRhTiAl (**a**) and ZrRhTiGa (**b**).

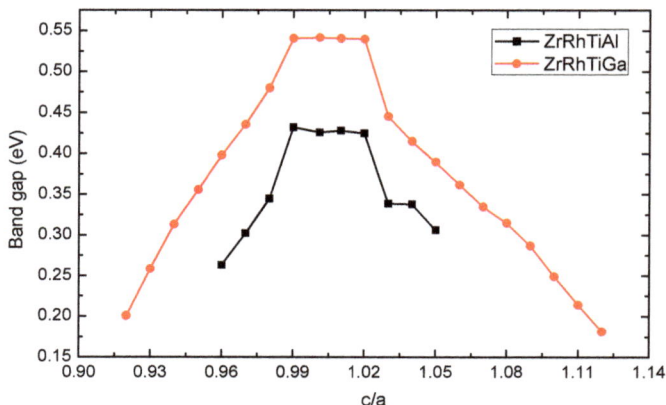

Figure 12. Band gaps as functions of the c/a ratio (TD) for ZrRhTiAl and ZrRhTiGa.

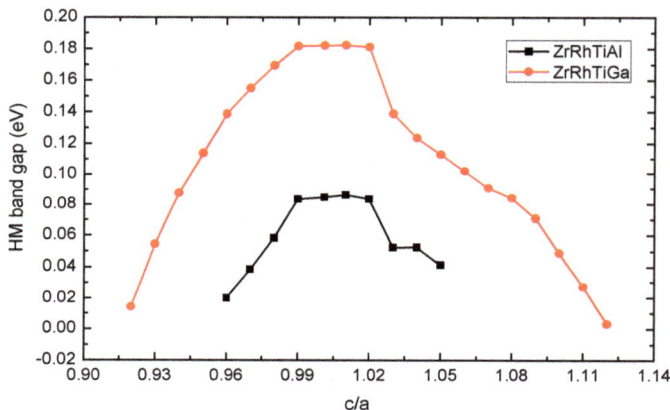

Figure 13. The half-metallic band-gaps as functions of the c/a ratio (TD) for ZrRhTiAl and ZrRhTiGa.

Next, the elastic properties were investigated for ZrRhTiZ (Z = Al, Ga) compounds by calculating the single crystal elastic constants C_{ij}. Heusler compounds belong to a cubic structure which needs only three independent elastic constants (C_{11}, C_{12}, and C_{44}) for the complete description of their elastic properties.

The following expressions are used to depict the elastic behavior of a material given in Reference [19]:

$$B = \frac{C_{11} + 2C_{12}}{3} \tag{1}$$

$$G = \frac{G_R + G_V}{2} \tag{2}$$

$$G_V = \frac{C_{11} - C_{12} + 3C_{44}}{5} \tag{3}$$

$$G_R = \frac{5(C_{11} - C_{12})C_{44}}{4C_{44} + 3(C_{11} - C_{12})} \tag{4}$$

$$E = \frac{9GB}{3B + G} \tag{5}$$

$$A = \frac{2C_{44}}{C_{11} - C_{12}} \tag{6}$$

where B, G, E, and A stands for bulk modulus, shear modulus, Young's modulus, and anisotropy factor, respectively. The estimated elastic constants and the mechanical properties for ZrRhTiAl and ZrRhTiGa are given in Table 3.

For cubic structures, the following criteria for the mechanical stability must be satisfied [20]:

$$C_{44} > 0 \tag{7}$$

$$\frac{(C_{11} - C_{12})}{2} > 0 \tag{8}$$

$$B > 0 \tag{9}$$

$$C_{12} < B < C_{11} \tag{10}$$

Therefore, from Table 3, one can easily see that ZrRhTiZ (Z = Al, Ga) compounds are mechanically stable. The calculated results of Young's modulus (E = 57.9 and 47.8 for ZrRhTiAl and ZrRhTiGa, respectively) indicate that ZrRhTiAl is stiffer than ZrRhTiGa. The Pugh's index (B/G) and Cauchy pressure ($C_{12} - C_{44}$) are usually used to characterize the brittleness or ductility of a compound [21–23]. For our materials, the Pugh's index (B/G) is 2.92 and 4.01, and the Cauchy pressures (C12 − C44) are 41.91 and 37.91 for the ZrRhTiAl and ZrRhTiGa compounds, respectively. Elastic anisotropy (A) is also a very important parameter for engineering applications. An A value of 1.0 shows the isotropic nature of the crystal, whereas values other than 1.0 indicate anisotropy. As shown in Table 3, for the ZrRhTiAl and ZrRhTiGa compounds, the value of A $\neq$ 1, demonstrating that they are elastically anisotropic under ambient conditions.

Finally, we have computed the formation and cohesive energies for ZrRhTiZ (Z = Al, Ga) based on the following expressions [24–28]:

$$E_f = E_{ZrRhTiZ}^{total} - (E_{Zr}^{bulk} + E_{Rh}^{bulk} + E_{Ti}^{bulk} + E_Z^{bulk}) \tag{11}$$

$$E_c = (E_{Zr}^{iso} + E_{Rh}^{iso} + E_{Ti}^{iso} + E_Z^{iso}) - E_{ZrRhTiZ}^{total} \tag{12}$$

The calculated E_c and E_f for ZrRhTiZ (Z = Al, Ga) compounds have been listed in Table 3. The E_c is larger than 20 eV, which indicates that ZrRhTiZ (Z = Al, Ga) compounds are stable due to energetic chemical bonding. The E_f is smaller than 0 eV, showing the compounds are stable against separation into the four pure elements.

Table 3. Calculated elastic constants C_{ij}, bulk modulus B, shear modulus G, Young's modulus E (GPa), Pugh's ratio B/G, anisotropy factor A, and formation and cohesive energies (eV) for the stoichiometric quaternary Heusler (SQH) compounds ZrRhTiAl and ZrRhTiGa.

SQH Compound	C_{11}	C_{12}	C_{44}	B	G	E	B/G	Formation Energy	Cohesive Energy	Anisotropy Factor
ZrRhTiAl	157.060	116.435	74.521	129.977	44.441	57.920	2.925	−2.632	22.279	3.669
ZrRhTiGa	150.857	117.550	79.645	128.652	32.073	47.895	4.011	−2.738	21.384	2.981

4. Summary

In this paper, the electronic structures, magnetic structures and moments, elastic constants, formation energies, and cohesive energies are calculated for ZrRhTiZ (Z = Al, Ga) quaternary Heusler compounds. Based on the obtained results, we predicted that ZrRhTiZ (Z = Al, Ga) quaternary Heusler compounds are half-metallic materials. These two compounds are among the rare SQH compounds with half-metallicity that contain 4d transition metal elements. The investigations on the effects of US and TD show that the half-metallicity is robust to US and TD. The half-metallicity can be sustained over

Appl. Sci. **2019**, *9*, 883

a wide lattice constant range of 6.09 Å–6.61 Å for ZrRhTiAl and 6.03 Å–6.75 Å for ZrRhTiGa. Under TD, the range of c/a ratio maintaining the half-metallicity is 0.96–1.06 for ZrRhTiAl and 0.92–1.12 for ZrRhTiGa, respectively. The calculated elastic parameters are listed in this paper for ZrRhTiZ (Z = Al, Ga) compounds to help understand their mechanical properties.

Author Contributions: Conceptualization, W.L. and G.L.; methodology, X.Z.; validation, H.J.; formal analysis, X.D.; investigation, W.L.; writing—original draft preparation, W.L.; writing—review and editing W.L., R.K. and G.L.; project administration, G.L.

Funding: This work was supported by the Natural Science Foundation of Tianjin City (No. 16JCYBJC17200) and the 333 Talent Project of Hebei Province (No. A2017002020).

Conflicts of Interest: The authors declare no conflict of interest.

References

1. Groot, R.A.D.; Mueller, F.M.; Engen, P.G.V.; Buschow, K.H.J. New class of materials—Half-metallic ferromagnets. *Phys. Rev. Lett.* **1983**, *50*, 2024–2027. [CrossRef]

2. Wurmehl, S.; Fecher, G.H.; Kandpal, H.C.; Ksenofontov, V.; Felser, C.; Lin, H.J. Investigation of Co_2FeSi: The Heusler compound with highest Curie temperature and magnetic moment. *Appl. Phys. Lett.* **2006**, *88*, 032503. [CrossRef]

3. Bainsla, L.; Mallick, A.I.; Raja, M.M.; Coelho, A.A.; Nigam, A.K.; Johnson, D.D.; Suresh, K.G. Origin of spin gapless semiconductor behavior in CoFeCrGa: Theory and Experiment. *Phys. Rev. B* **2015**, *92*, 045201. [CrossRef]

4. Guo, R.K.; Liu, G.D.; Wang, X.T.; Rozale, H.; Wang, L.Y.; Khenata, R.; Wu, Z.M.; Dai, X.F. First-principles study on quaternary Heusler compounds ZrFeVZ (Z = Al, Ga, In) with large spin-flip gap. *RSC Adv.* **2016**, *6*, 109394–109400. [CrossRef]

5. Wang, X.T.; Cheng, Z.X.; Wang, J.L.; Wang, L.Y.; Yu, Z.Y.; Fang, C.S.; Yang, J.T.; Liu, G.D. Origin of the half-metallic band-gap in newly designed quaternary Heusler compounds ZrVTiZ (Z = Al, Ga). *RSC Adv.* **2016**, *6*, 57041–57047. [CrossRef]

6. Berri, S.; Ibrir, M.; Maouche, D.; Attallah, M. Robust half-metallic ferromagnet of quaternary Heusler compounds ZrCoTiZ (Z = Si, Ge, Ga and Al). *Comput. Conden. Matter.* **2014**, *1*, 26–31. [CrossRef]

7. Berri, S.; Ibrir, M.; Maouche, D.; Attallah, M. First principles study of structural, electronic and magnetic properties of ZrFeTiAl, ZrFeTiSi, ZrFeTiGe and ZrNiTiAl. *J. Magn. Magn. Mater.* **2014**, *371*, 106–111. [CrossRef]

8. Wang, J.X.; Chen, Z.B.; Gao, Y.C. Phase stability, magnetic, electronic, half-metallic and mechanical properties of a new equiatomic quaternary Heusler compound ZrRhTiIn: A first-principles investigation. *J. Phys. Chem. Solids.* **2018**, *116*, 72–78. [CrossRef]

9. Wang, X.T.; Zhao, W.Q.; Cheng, Z.X.; Dai, X.F.; Khenata, R. Electronic, magnetic, half-metallic and mechanical properties of a new quaternary Heusler compound ZrRhTiTl: Insights from first-principles studies. *Solid State Commun.* **2018**, *269*, 125–130. [CrossRef]

10. Segall, M.D.; Lindan, P.L.D.; Probert, M.J.; Pickard, C.J.; Hasnip, P.J.; Clark, S.J.; Payne, M.C. First-principles simulation: Ideas, illustrations and the CASTEP code. *J. Phys. Condens. Matter.* **2002**, *14*, 2717. [CrossRef]

11. Payne, M.C.; Teter, M.P.; Allan, D.C.; Arias, T.A.; Joannopoulos, J.D. Iterative Minimization Techniques for ab Initio Total-Energy Calculations: Molecular Dynamics and Conjugate Gradients. *Rev. Mod. Phys.* **1992**, *64*, 1045. [CrossRef]

12. Perdew, J.P.; Chevary, J.A.; Vosko, S.H.; Jackson, K.A.; Pederson, M.R.; Singh, D.J.; Fiolhais, C. Atoms, Molecules, Solids, and Surfaces: Applications of the Generalized Gradient Approximation for Exchange and Correlation. *Phys. Rev. B* **1992**, *46*, 6671. [CrossRef]

13. Perdew, J.P.; Burke, K.; Ernzerhof, M. Generalized Gradient Approximation Made Simple. *Phys. Rev. Lett.* **1996**, *77*, 3865. [CrossRef] [PubMed]

14. Wang, X.T.; Lin, T.T.; Rozale, H.; Dai, X.F.; Liu, G.D. Robust half-metallic properties in inverse Heusler alloys composed of 4d transition metal elements: Zr_2RhZ (Z = Al, Ga, In). *J. Magn. Magn. Mater.* **2016**, *402*, 190–195. [CrossRef]

15. Zhang, L.; Wang, X.T.; Rozale, H.; Lu, J.W.; Wang, L.Y. Half-Metallicity and Tetragonal Deformation of Ti$_2$RhAl, Ti$_2$RhGa, and Ti$_2$RhIn: A First-Principle Study. *J. Supercond. Nov. Magn.* **2016**, *29*, 349–356. [CrossRef]

16. Kang, X.H.; Zhang, J.M. The structural, electronic and magnetic properties of a novel quaternary Heusler alloy TiZrCoSn. *J. Phys. Chem. Solids* **2017**, *105*, 9–15. [CrossRef]

17. Yan, P.L.; Zhang, J.M.; Xu, K.W. The structural, electronic and magnetic properties of quaternary Heusler alloy TiZrCoIn. *Solid State Commun.* **2016**, *231*, 64–67. [CrossRef]

18. Berri, S. Electronic structure and half-metallicity of the new Heusler alloys PtZrTiAl, PdZrTiAl and Pt$_{0.5}$Pd$_{0.5}$ZrTiAl. *Chin. J. Phys.* **2017**, *55*, 195–202. [CrossRef]

19. Ahmad, M.; Murtaza, G.; Khenata, R.; Omran, S.B.; Bouhemadou, A. Structural, elastic, electronic, magnetic and optical properties of RbSrX(C, SI, Ge) half-Heusler compounds. *J. Magn. Magn. Mater.* **2015**, *377*, 204–210. [CrossRef]

20. Wang, X.T.; Cheng, Z.X.; Wang, J.L.; Liu, G.D. A full spectrum of spintronic properties demonstrated by a C1b-type Heusler compound Mn$_2$Sn subjected to strain engineering. *J. Mater. Chem. C* **2016**, *4*, 8535–8544. [CrossRef]

21. Pugh, S.F. XCII. Relations between the elastic moduli and the plastic properties of polycrystalline pure metals. *Lond. Edinb. Dublin Philos. Mag. J. Sci.* **1954**, *45*, 823–843. [CrossRef]

22. Bhat, T.M.; Gupta, D.C. Effect of on-site Coulomb interaction on electronic and transport properties of 100% spin polarized CoMnVAs. *J. Magn. Magn. Mater.* **2017**, *435*, 173–178. [CrossRef]

23. Jain, R.; Jain, V.K.; Chandra, A.R.; Jain, V.; Lakshmi, N. Study of the Electronic Structure, Magnetic and Elastic Properties and Half-Metallic Stability on Variation of Lattice Constants for CoFeCrZ (Z = P, As, Sb) Heusler Alloys. *J. Supercond. Nov. Magn.* **2018**, *31*, 2399. [CrossRef]

24. Zhang, L.; Gao, Y.C. Electronic structures, magnetic properties and half-metallicity in the Heusler alloy Hf2Val. *Chin. J. Phys.* **2017**, *55*, 1466. [CrossRef]

25. Wang, X.T.; Cheng, Z.X.; Wang, J.L.; Rozale, H.; Wang, L.; Yu, Z.Y.; Liu, G.D. Strain-induced diverse transitions in physical nature in the newly designed inverse Heusler alloy Zr$_2$MnAl. *J. Alloys Compd.* **2016**, *686*, 549–555. [CrossRef]

26. Luo, H.Z.; Xin, Y.P.; Liu, B.H.; Meng, F.B.; Liu, H.Y.; Liu, E.K.; Wu, G.H. Competition of L21 and XA structural ordering in Heusler alloys X2CuAl (X = Sc, Ti, V, Cr, Mn, Fe, Co, Ni). *J. Alloys Compd.* **2016**, *665*, 180–185. [CrossRef]

27. Zhao, J.S.; Gao, Q.; Li, L.; Xie, H.H.; Hu, X.R.; Xu, C.L.; Deng, J.B. First-principles study of the structure, electronic, magnetic and elastic properties of half-Heusler compounds LiXGe (X = Ca, Sr and Ba). *Intermetallics* **2017**, *89*, 65–73. [CrossRef]

28. Birsan, A. Small interfacial distortions lead to significant changes of the half-metallic and magnetic properties in Heusler alloys: The case of the new CoFeZrSi compound. *J. Alloys Compd.* **2017**, *710*, 393–398. [CrossRef]

Article

First-Principles Investigation of Atomic Hydrogen Adsorption and Diffusion on/into Mo-doped Nb (100) Surface

Yang Wu [1], **Zhongmin Wang** [1,2,*], **Dianhui Wang** [1,*] ![ORCID], **Jiayao Qin** [1], **Zhenzhen Wan** [1], **Yan Zhong** [1,2], **Chaohao Hu** [1,2,*] and **Huaiying Zhou** [1,2]

[1] School of Materials Science and Engineering, Guilin University of Electronic Technology, Guilin 541004, China; wuyang@ihep.ac.cn (Y.W.); qjyhsm2012@163.com (J.Q.); qingfeng521567@163.com (Z.W.); zyguet@163.com (Y.Z.); zhy@guet.edu.cn (H.Z.)

[2] Guangxi Key Laboratory of Information Materials, Guilin University of Electronic Technology, Guilin 541004, China

* Correspondence: zmwang@guet.edu.cn (Z.W.); devix@mails.guet.edu.cn (D.W.); chaohao.hu@guet.edu.cn (C.H.); Tel.: +86-139-7831-6492 (Z.W.); +86-150-0773-1239 (D.W.); +86-137-6843-4525 (C.H.)

Received: 6 November 2018; Accepted: 28 November 2018; Published: 3 December 2018

Abstract: To investigate Mo doping effects on the hydrogen permeation performance of Nb membranes, we study the most likely process of atomic hydrogen adsorption and diffusion on/into Mo-doped Nb (100) surface/subsurface (in the $Nb_{12}Mo_4$ case) via first-principles calculations. Our results reveal that the (100) surface is the most stable Mo-doped Nb surface with the smallest surface energy ($2.75 \, J/m^2$). Hollow sites (HSs) in the Mo-doped Nb (100) surface are H-adsorption-favorable mainly due to their large adsorption energy (-4.27 eV), and the H-diffusion path should preferentially be HS→TIS (tetrahedral interstitial site) over HS→OIS (octahedral interstitial site) because of the correspondingly lower H-diffusion energy barrier. With respect to a pure Nb (100) surface, the Mo-doped Nb (100) surface has a smaller energy barrier along the HS→TIS pathway (0.31 eV).

Keywords: Nb (100) surface; Mo doping; H adsorption; H diffusion; first-principles calculation

1. Introduction

The interaction between hydrogen and metal surfaces is an interesting topic in science and engineering, and has been investigated by both experimental [1–3] and theoretical [4,5] approaches. Over 80 percent of synthetic chemicals are produced with the application of catalysts, meaning that the interactions of hydrogen with catalytic metal surfaces during heterogeneous catalysis are of great interest for several important processes [6–8] including petrochemical processing, pharmaceutical production [9], highly efficient electrocatalysis [10], fine chemical production [11], and conversion of biomass to fuels and chemicals [12]. Meanwhile, hydrogen–metal interactions have attracted further attention due to the central role of hydrogen as a clean and efficient energy source [13,14]. With the increasing demand for high purity hydrogen in the fields of fuel cell, new materials for hydrogen separation/purification are being explored. Consequently, the study of the behavior of both absorbed and adsorbed hydrogen on/into the metal surface/subsurface is important for achieving a deep understanding of the hydrogen-permeation properties of such metallic membranes [15,16].

Considerable experimental and theoretical studies of the hydrogen sorption on many single-crystal metal surfaces have been carried out. For example, Ferrin et al. [4,15] performed a comprehensive theoretical study addressing adsorption and absorption energies and subsurface penetration barriers of hydrogen on different low-index surface terminations of 17 transition metals (TM) by first-principles calculations. Lauhon et al. further studied the diffusion of atomic hydrogen on Cu (001) via scanning tunneling microscope (STM) measurements. The diffusion of H atoms was measured as a function of temperature, and a transition from thermally activated diffusion to quantum tunneling was observed at 60 K [16]. Recently, Gómez et al. reported an extensive study of adsorption and diffusion of hydrogen atoms on the (100) surfaces of fcc Au, Cu, Ag, and Pt, performed by means of density functional theory (DFT) calculations [17,18]. Based on elucidating the preferential adsorption sites of hydrogen on each metal, they calculated the adsorption distances and energies of atomic hydrogen at the top, hollow, and bridge sites of the (100) surfaces. The numerous studies focused on this topic have proved the importance of understanding the mechanism of hydrogen–metal-surface-related phenomena.

Pd and its alloys are excellent materials for hydrogen separation and purification, but their high cost and scarcity should be noted. Currently, niobium (Nb), a Group VB (e.g., V, Nb, Ta, etc.) TM, has attracted much attention as one of the most promising hydrogen separation materials due to its relatively low price [3], excellent high-temperature mechanical properties, and corrosion resistance [19–21]. However, Nb often exhibit poor resistance to hydrogen embrittlement and therefore is limited in the practical application [22–25]. Experimental and theoretical studies have verified that alloying Nb with other metals is an effective approach to solve this problem [26–30]. Hu et al. [31] reported that the W doping is supposed to be the key role in enhancing the mechanical properties of $Nb_{16}H$, and is not conducive to the structural stability of the $Nb_{15}WH$ (TIS) phase. Moreover, doping Nb with W can reduce the diffusion barrier of H, and enhance diffusion paths for H [32].

Both W and Mo are high-Z refractory metals (i.e., refractory metals containing impurities with high atomic numbers (Z) with similar physical properties. Comparing with W, Mo has a lower melting point (2883 K) and a lower erosion rate. Moreover, H has higher diffusivity and lower solubility in Mo, leading to lower H retention [33–35]. These characteristics make Mo an important alloying candidate of Nb-based membranes for hydrogen permeation.

To gain a detailed understanding of the hydrogen-permeation behavior of Mo-doped Nb membranes, adsorption and diffusion of hydrogen atoms on a Mo-doped Nb ($Nb_{12}Mo_4$) surface have been investigated by first-principles calculations in this work. Furthermore, we discuss the most favorable process of H adsorption and diffusion on/into a Mo-doped Nb (100) surface/subsurface. We believe this work is important to comprehensively understand the basic mechanism of atomic hydrogen sorption on TM surfaces and the influence of element doping, and contribute to the design of Nb-based alloys for H-storage and H-separation applications.

2. Computational Details

Our calculations were carried out with the use of the Vienna Ab-initio Simulation Package (VASP) [36,37]. The interactions between the core and valence electrons were described with the projector augmented wave (PAW) approach [38,39]. Exchange correlation functions were generalized gradient approximations (GGA) developed by Perdew et al. [40]. An energy cutoff of 360 eV was used for the plane-wave basis sets, and a grid with $2\pi \times 0.03$ Å^{-1} resolution in the Brillouin zone was used for all calculations to minimize the error from the k-point meshes. During structure relaxation, the lattice parameters, volume, and atomic positions were fully optimized with in-symmetry restrictions until the total energy converged to 10^{-5} eV in the self-consistent loop. Relaxed until the maximum force on each atom was less than 0.01 eV/Å. The electrons of the Nb $4d^45s^1$ and the Mo $4d^55s^1$ orbitals were treated as valence electrons. To study the diffusion properties of atomic hydrogen from the surface into the subsurface of Mo-doped Nb (100), we used the climbing image nudged elastic band (CI-NEB) method [41] to determine the diffusion barriers between the initial and final positions.

3. Results and Discussion

3.1. Surface Model of Mo-doped Nb

In $Nb_{16-x}Mo_xH$ system, $Nb_{12}Mo_4H$ has the highest absolute value of ΔH_f, which is favorable for dehydrogenation. Therefore, $Nb_{12}Mo_4H$ was selected to build the slab model for further study, which is depicted in Figure 1a. There are two high symmetrical interstitial sites in the region between the first and second atomic layer, which are tetrahedral interstitial site (TIS) and octahedral interstitial site (OIS), as shown in Figures 1 and 1. Due to the similar atomic radius of Mo compared to Nb, the crystal keeps the cubic structure after Mo-doping. The equilibrium lattice constant (3.304 Å) of bulk Nb agrees well with experimental values (3.305 Å) [42]. For the (100) surface, we built a slab model with seven atomic layers, which is depicted in Figure 1d.

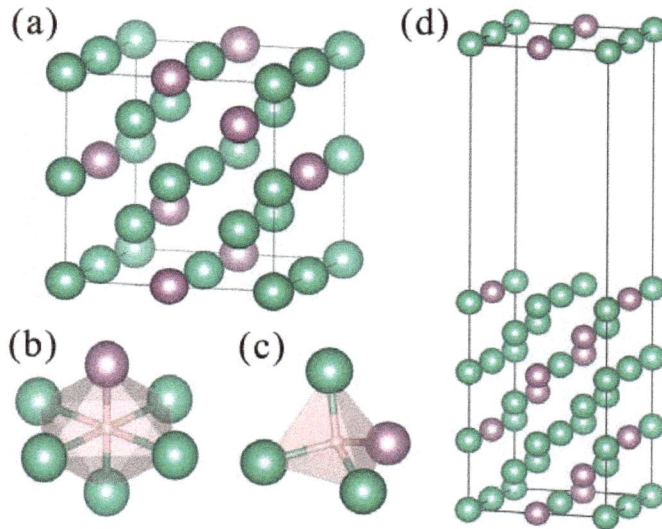

Figure 1. Theoretical models of $Nb_{12}Mo_4$ used in the calculations: (**a**) bulk model; (**b**) an octahedral interstitial site (OIS), (**c**) a tetrahedral interstitial site (TIS); (**d**) slab model. Nb atoms are represented by green spheres, and Mo atoms are represented by gray spheres.

Surface energy is an important parameter of a metal surface that can be utilized to explain the physical and chemical processes and surface stability. Cutting surfaces from bulk crystal breaking various chemical bonds, which increases the energy of the system, thereby resulting in instability of the surface systems. The surface energy is defined as [14]

$$\sigma = (E_{slab} - n \times E_{bulk})/(2A) \tag{1}$$

Here, E_{slab}, E_{bulk}, n, and A represent the total energy of the slab, the total energy of the corresponding bulk material, the number of atoms contained in the slab and the area of the surface, respectively. In the structure optimization calculations, the atoms in the top three layers were fully relaxed, and the atoms in the other layers were fixed at their bulk positions. Table 1 lists the calculated surface energies of slab models with different numbers of atomic layers. The surface energy of pure Nb (100) surface is also calculated for comparison. The results agree well with the values from Baskes et al. [43]. The surface energy of $Nb_{12}Mo_4(100)$ surface and $Nb_{12}Mo_4(111)$ surface tend to be stable when there are more than five atomic layers. Moreover, the $Nb_{12}Mo_4(100)$ surface is the most stable one with the smallest surface energy (2.75 J/m^2). Therefore, we select Nb$_{12}$Mo$_4$(100) surface for further study.

Table 1. Calculated surface energies of pure Nb (100), Nb$_{12}$Mo$_4$ (100) and (111) for of different numbers of atomic layers. Experimental value [43] is listed for comparison.

	Surface Energies (eV)						
	4 Layers	**5 Layers**	**6 Layers**	**7 Layers**	**8 Layers**	**9 Layers**	**10 Layers**
Pure Nb (100)	2.63	2.45	2.43	2.42	2.44	2.44	2.44
Pure Nb (100) [a]		2.3–2.7					
Nb$_{12}$Mo$_4$ (100)	2.72	2.73	2.74	2.75	2.75	2.75	2.75
Nb$_{12}$Mo$_4$ (111)	3.28	3.31	3.32	3.32	3.32	3.32	3.32

[a] Ref. [43].

To determine a practicable number of atomic layers and the vacuum region of Mo-doped Nb (100) surface model, we calculate the changes of the interlayer distances during structure relaxation as the following relationship [44,45]

$$\Delta d_{i-j} = d_{i-j} - d_0 \tag{2}$$

Here, d_0 denotes the layer distance between the i- and j-layers of the unrelaxed slab model established from a geometry-optimized bulk model and d_{i-j} denotes the corresponding layer distance of the optimized slab model. From the calculated results depicted in Figure 2, we can infer that the interlayer distance between the outermost layer and secondary outer layer tends to be stable when there are more than six atomic layers. According to the previously calculated surface energies, seven atomic layers are sufficient to avoid the effects of external forces in z-axis.

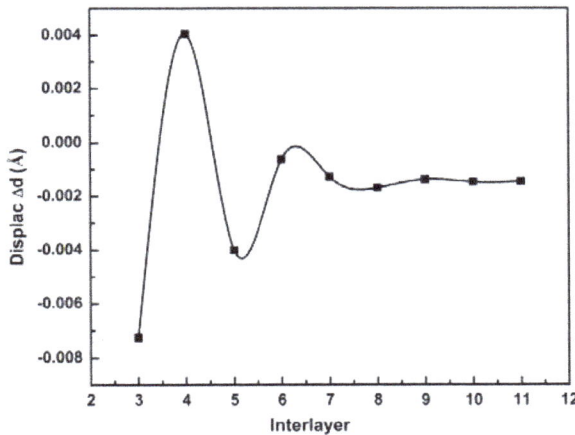

Figure 2. Change in interlayer distance as function of interlayer number in Nb (100) surface model.

In addition, to ensure that hydrogen permeation process is not influenced by periodicity, a vacuum region between the atomic layers in the (100) surface is required. Thus, the work function of (100) surface model with seven atomic layers and a 12 Å vacuum layer along the surface normal direction (z-axis) is calculated, and is depicted in Figure 3. The flat line in the figure indicating that the work function converges in the vacuum region. Therefore, we employed a slab model of seven atomic layers with a 12 Å vacuum layer of (100) surface for further hydrogen permeation analysis.

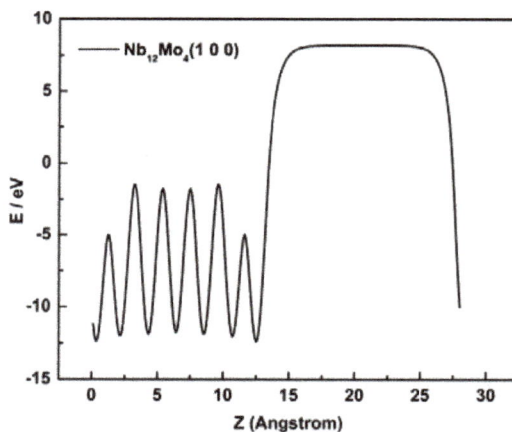

Figure 3. Work function along surface normal direction (Z-axis) for Nb$_{12}$Mo$_4$ (100) surface model with seven atomic layers.

3.2. H Atom Adsorption Sites of Mo-doped Nb (100) Surface

Based on the slab model of Mo-doped Nb (100) surface, there are three possible adsorption sites for hydrogen atoms, i.e., top site (TS), bridge site (BS), and hollow site (HS), as shown in Figure 4. The adsorption energies (E_{ads}) are calculated to characterize the preferred location of the adsorbed hydrogen atom. When Nb doping is considered, E_{ads} can be expressed as

$$E_{ads} = E(Nb_nMo_{m-n}H) - E(Nb_nMo_{m-n}) - E(H) \qquad (3)$$

Here, m represents the total number of atoms, n represents the number of Nb atoms in the slab model, $E(H) = -0.019$ eV, $E(Nb_nMo_{m-n})$ and $E(Nb_nMo_{m-n}H)$ the total energies of the corresponding crystal structures. The calculated E_{ads} values are summarized in Table 2. Among the possible positions, the largest adsorption energy (-4.27 eV) belongs to the hollow site on top of Nb atom (H1), which indicates that H1 sites are the preferential sites of adsorbed hydrogen atoms. In addition, the average vertical distance between adsorbed H atom and the top layer ($d_{\text{H-Surf}}$) as well as the distance between H atom and TM atoms ($d_{\text{H-TM}}$) are also calculated and listed in Table 2. The bonding lengths ($d_{\text{H-TM}}$) between the hydrogen-atom and TM atom show the descending on the hollow, bridge and top sites, respectively. The smallest $d_{\text{H-Surf}}$ (0.44 Å) belongs to the H1 site, indicating that there exists a very strong interaction between H atom and (100) surface. Similar results were obtained by Gong and co-authors [46].

Table 2. Calculated adsorption energy (E_{ads}), average distance between H atom and metal atom ($d_{\text{H-TM}}$) as well as average distance between H atom and the top layer ($d_{\text{H-Surf}}$) in slab model of Mo-doped Nb (100) surface for adsorbed hydrogen atoms at different sites indicated in Figure 4

Site	E_{ads} (eV)	$d_{\text{H-TM}}$ (Å)	$d_{\text{H-Surf}}$ (Å)
T2	-3.41	1.77	1.85
T1	-3.26	1.85	1.93
B2	-3.77	1.97	1.30
B1	-3.75	1.97	1.21
H2	-4.24	1.98	0.46
H1	-4.27	1.99	0.44

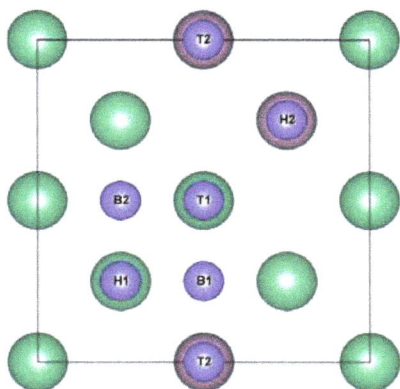

Figure 4. Schematic illustration of three possible locations of hydrogen atoms adsorbed onto Mo-doped Nb (100) surface: top site (T1, T2), bridge site (B1, B2), and hollow site (H1, H2).

3.3. Electronic Properties of Mo-doped Nb (100) Surface

To examine the interaction between atoms, we investigated the electronic properties of the Mo-doped Nb (100) surface. The calculated density of states (DOS) of H atom adsorbed at the bridge, hollow, and top sites are shown in Figure 5a–c, respectively. The Partial DOS (PDOS) peaks at about -30 eV with respect to the Fermi level are mainly contributed by Nb-p and Mo-p states for Mo-doped Nb (100) surface. The PDOS peaks near -5 eV are contributed by the H-s state. In the case of H1 model (Figure 5b), an obvious peak of Nb-d is observed at -5 eV, but it is absent in the case of B1 and T1 models, indicating that the hybridization between H-s and Nb-d states in B1 model is adjusted by Mo-doping.

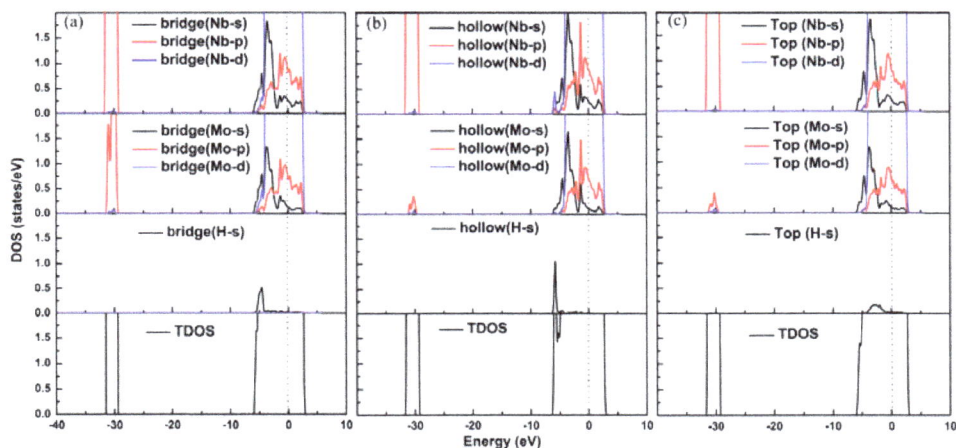

Figure 5. Calculated total and partial density of states of hydrogen atoms adsorbed onto Mo-doped Nb (100) surface at (**a**) B1 site, (**b**) H1 site, and (**c**) T1 site. The Fermi level is set to zero.

3.4. Diffusion of Hydrogen Atom into Mo-doped Nb (100) Subsurface

To further understand the behavior of H atoms diffusion, we investigate the subsequent H-diffusion process from the surface into the subsurface. The possible H-diffusion paths are H1→TIS and H1→OIS. We utilize CI-NEB method to determine the minimum energy and energy barrier for H-diffusion at both paths. The energy barrier of H1→TIS is 0.31 eV (Figure 6b), which is smaller than the corresponding value (1.0 eV) of H1→OIS (Figure 6a), meaning that the H-diffusion path should mainly be H1→TIS. In addition, the energy barrier of H1→TIS of the pure Nb (100) surface is 0.82 eV (Figure 6c), indicating Mo-doping improves the performance of H-atom diffusion for Nb (100) surface.

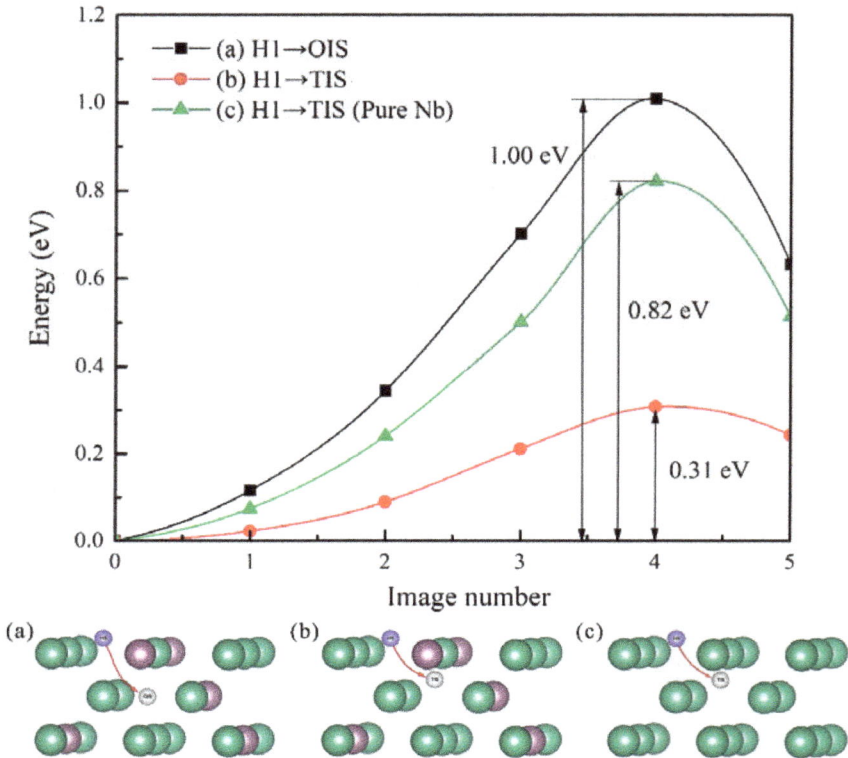

Figure 6. Energy barriers of H-atom diffusion in different path. (**a**) H1 to OIS in Mo-doped Nb (100) surface, (**b**) H1 to tetrahedral interstitial site in Mo-doped Nb (100) surface, and (**c**) HS to TIS in pure Nb (100) surface. Blue balls represent the starting position and white balls represent the finish position for H atoms.

4. Conclusions

The probable processes of atomic hydrogen adsorption, and diffusion behaviors on/into Mo-doped Nb (100) surface/subsurface are investigated using first-principles calculations in combination with empirical theory. Considering the above mentioned, we can come to the following conclusion:

(1) Comparing with the (111) surface, the (100) surface is more stable with smaller surface energy (2.75 J/m^2) for Mo-doped Nb.

(2) Among the three possible sites for H adsorption (BS, HS, and TS), H1 is most favorable due to its larger adsorption energy (−4.27 eV) and smaller average vertical distance between H atoms and the top layer of surface (0.44 Å). The optimal H-diffusion path is H1→TIS for its low energy barrier of H-diffusion.

(3) The optimal H-diffusion path is H1→TIS. Furthermore, compared with pure Nb, Mo-doped Nb (100) surface offers a smaller energy barrier for H1→TIS path (0.31 eV), indicating that Mo-doping improves H adsorption and diffusion performance for Nb (100) surface.

Author Contributions: Z.W. (Zhongmin Wang) and C.H. conceived the work; Z.W. (Zhenzhen Wan) and Y.W. performed the calculations; Y.W. and D.W. wrote this manuscript; J.Q., Y.Z. and H.Z. made valuable comments on this manuscript.

Funding: This work was financially supported by the National Natural Science Foundation of China (51471055, 11464008, 51401060, and 51761007), the Natural Foundations of Guangxi Province (2014GXNSFGA118001 and 2016GXNSFGA380001), Guangxi Key Laboratory of Information Materials (131022-Z), and Guangxi Experiment Center of Information Science (YB1512).

Conflicts of Interest: The authors declare no conflict of interest.

References

1. Ding, L.; Wei, Y.; Li, L.; Zhang, T.; Wang, H.; Xue, J.; Wang, S.; Caro, J.; Gogotsi, Y. MXene molecular sieving membranes for highly efficient gas separation. *Nat. Commun.* **2018**, *9*, 155. [CrossRef] [PubMed]

2. Wang, Z.L.; Hao, X.F.; Jiang, Z.; Sun, X.P.; Xu, D.; Wang, J.; Zhong, H.; Meng, F.; Zhang, X. C and N hybrid coordination derived Co-C-N complex as a highly efficient electrocatalyst for hydrogen evolution reaction. *J. Am. Chem. Soc.* **2015**, *48*, 15070–15073. [CrossRef] [PubMed]

3. Li, F.; Zhong, B.; Xiao, H.; Ye, X.; Lu, L.; Guan, W.; Zhang, Y.; Wang, X.; Chen, C. Effect of degassing treatment on the deuterium permeability of Pd-Nb-Pd composite membranes during deuterium permeation. *Sep. Purif. Technol.* **2018**, *190*, 136–142. [CrossRef]

4. Ferrin, P.; Kandoia, S.; Nilekara, A.U.; Mavrikakisa, M. Hydrogen adsorption, absorption and diffusion on and intransition metal surfaces: A DFT study. *Surf. Sci.* **2012**, *606*, 679–689. [CrossRef]

5. Wang, J.W.; He, Y.H.; Gong, H.R. Various properties of Pd$_3$Ag/TiAl membranes from density functional theory. *J. Membr. Sci.* **2015**, *475*, 406–413. [CrossRef]

6. Wang, J.; Zhong, H.; Wang, Z.; Meng, F.L.; Zhang, X.B. Integrated three-dimensional carbon paper/carbon tubes/cobalt-sulfide sheets as an efficient electrode for overall water splitting. *ACS Nano* **2016**, *2*, 2342–2348. [CrossRef] [PubMed]

7. Zhong, H.; Wang, J.; Meng, F.; Zhang, X. In Situ Activating Ubiquitous Rust towards Low-Cost, Efficient, Free-Standing, and Recoverable Oxygen Evolution Electrodes. *Angew. Chem. Int. Ed.* **2016**, *34*, 9937–9941. [CrossRef]

8. Singh, M.E.; Rezac, P.H.; Pfromm, P. Hydrogenation of soybean oil using metal decorated integral-asymmetric polymer membranes: Effects of morphology and membrane properties. *J. Membr. Sci.* **2010**, *348*, 99–108. [CrossRef]

9. Singh, U.K.; Vannice, M.A. Kinetics of liquid-phase hydrogenation reactions over supported metal catalysts—A review. *Appl. Catal. A* **2001**, *213*, 1–24. [CrossRef]

10. Wang, J.; Li, K.; Zhong, H.; Xu, D.; Wang, Z.L.; Jiang, Z.; Wu, Z.; Zhang, X.B. Synergistic effect between metal-nitrogen-carbon sheets and NiO nanoparticles for enhanced electrochemical water-oxidation performance. *Angew. Chem. Int. Ed.* **2015**, *36*, 10530–10534. [CrossRef]

11. Zacher, A.H.; Olarte, M.V.; Santosa, D.M.; Elliott, D.C.; Jones, S.B. A review and perspective of recent bio-oil hydrotreating research. *Green Chem.* **2014**, *16*, 491–515. [CrossRef]

12. Carrette, L.; Friedrich, K.A.; Stimming, U. Fuel cells: Principles, types, fuels, and applications. *J. Chem. Phys.* **2000**, 1162–1193. [CrossRef]

13. Edwards, P.P.; Kuznetsov, V.L.; David, W.I.F.; Brandon, N.P. Hydrogen and fuel cells: towards a sustainable energy future. *Energy Policy* **2008**, *36*, 4356–4362. [CrossRef]

14. Kikuchi, E. Membrane reactor application to hydrogen production. *Catal. Today* **2000**, *56*, 97–101. [CrossRef]

15. Dong, W.; Ledentu, V.; Sautet, P.; Eichler, A.; Hafner, J. Hydrogen adsorption on palladium: A comparative theoretical study of different surfaces. *Surf. Sci.* **1998**, *411*, 123–136. [CrossRef]

16. Lauhon, L.J.; Ho, W. Direct observation of the quantum tunneling of single hydrogen atoms with a scanning tunneling microscope. *Phys. Rev. Lett.* **2000**, *85*, 4566. [CrossRef] [PubMed]

17. Gómez, E.D.V.; Amayaroncancio, S.; Avalle, L.B.; Linares, D.H.; Gimenez, M.C. DFT study of adsorption and diffusion of atomic hydrogen on metal surfaces. *Appl. Surf. Sci.* **2017**, *420*, 1–8. [CrossRef]

18. Mitsui, T.; Rose, M.K.; Fomin, E.; Ogletree, D.F.; Salmeron, M. Hydrogen adsorption and diffusion on Pd(111). *Surf. Sci.* **2003**, *540*, 5–11. [CrossRef]

19. Liu, Z.H.; Shang, J.X. Elastic properties of Nb-based alloys by using the density functional theory. *Chin. Phys. B* **2012**, *21*, 016202. [CrossRef]

20. Liu, G.; Besedin, S.; Irodova, A.; Liu, H.; Gao, G.; Eremets, M.; Wang, X.; Ma, Y. Nb-H system at high pressures and temperatures. *Phys. Rev. B* **2017**, *95*, 104110. [CrossRef]

21. Peterson, D.T.; Hull, A.B.; Loomis, B.A. Hydrogen embrittlement considerations in niobium-base alloys for application in the ITER divertor. *J. Nucl. Mater.* **2012**, *191–194*, 430–432.

22. Watanabe, N.; Yukawa, H.; Nambu, T.; Matsumoto, Y.; Zhang, G.X.; Morinaga, M. Alloying effects of Ru and W on the resistance to hydrogen embrittlement and hydrogen permeability of niobium. *J. Alloys Compd.* **2009**, *477*, 851–854. [CrossRef]

23. Zhang, G.X.; Yukawa, H.; Nambu, T.; Matsumoto, Y.; Morinaga, M. Alloying effects of Ru and W on hydrogen diffusivity during hydrogen permeation through Nb-based hydrogen permeable membranes. *Int. J. Hydrogen Energy* **2010**, *35*, 1245–1249. [CrossRef]

24. Yukawa, H.; Nambu, T.; Matsumoto, Y.; Watanabe, N.; Zhang, G.; Morinaga, M. Alloy design of Nb-based hydrogen permeable membrane with strong resistance to hydrogen embrittlement. *Mater Trans.* **2008**, *49*, 2202–2207. [CrossRef]

25. Nambu, T.; Shimizu, K.; Matsumoto, Y.; Rong, R.; Watanabe, N.; Yukawa, H.; Morinaga, M.; Yasuda, I. Enhanced hydrogen embrittlement of Pd-coated niobium metal membrane detected by in situ small punch test under hydrogen permeation. *J. Alloys Compd.* **2007**, *446*, 588–592. [CrossRef]

26. Long, J.H.; Gong, H. Phase stability and mechanical properties of niobium dihydride. *Int. J. Hydrogen Energy* **2014**, *39*, 18989–18996. [CrossRef]

27. Slining, J.R.; Koss, D.A. Solid solution strengthening of high purity niobium alloys. *Metall. Trans.* **1973**, *4*, 1261–1264. [CrossRef]

28. Kozhakhmetov, S.; Sidorov, N.; Piven, V.; Sipatov, I.; Gabis, I.; Arinov, B. Alloys based on Group 5 metals for hydrogen purification membranes. *J. Alloys Comp.* **2015**, *645*, S36–S40. [CrossRef]

29. Xu, Z.J.; Wang, Z.M.; Tang, J.L.; Deng, J.Q.; Yao, Q.R.; Zhou, H.Y. Effects of Mo alloying on the structure and hydrogen-permeation properties of Nb metal. *J. Alloys Compd.* **2018**, *740*, 810–815. [CrossRef]

30. Wu, Y.; Wang, Z.M.; Wang, D.H.; Wan, Z.Z.; Zhong, Y.; Hu, C.H.; Zhou, H. Effects of Ni doping on various properties of NbH phases: A first-principles investigation. *Sci. Rep. UK* **2017**, *7*, 6535. [CrossRef]

31. Hu, Y.T.; Gong, H.; Chen, L. Fundamental effects of W alloying on various properties of NbH phases. *Int. J. Hydrogen Energy* **2015**, *40*, 12745–12749. [CrossRef]

32. Kong, X.S.; Wang, S.; Wu, X.; You, Y.W.; Liu, C.S.; Fang, Q.F.; Chen, J.L.; Luo, G.N. First-principles calculations of hydrogen solution and diffusion in tungsten: Temperature and defect-trapping effects. *Acta Mater.* **2015**, *84*, 426–435. [CrossRef]

33. Nagata, S.; Takahiro, K. Deuterium retention in tungsten and molybdenum. *J. Nucl. Mater.* **2000**, *283*, 1038–1042. [CrossRef]

34. Duan, C.; Liu, Y.L.; Zhou, H.B.; Zhang, Y.; Jin, S.; Lu, G.H.; Luo, G.N. First-principles study on dissolution and diffusion properties of hydrogen in molybdenum. *J. Nucl. Mater.* **2010**, *404*, 109–115. [CrossRef]

35. Wright, G.M.; Whyte, D.G.; Lipschultz, B. Measurement of hydrogenic retention and release in molybdenum with the DIONISOS experiment. *J. Nucl. Mater.* **2009**, *390–391*, 544–549. [CrossRef]

36. Kresse, G.; Hafner, J. Ab-initio molecular dynamics for liquid metals. *Phys. Rev. B* **1993**, *47*, 558–561. [CrossRef]

37. Kresse, G.; Furthmüller, J. Efficient iterative schemes for ab initio total-energy calculations using a plane-wave basis set. *Phys. Rev. B* **1996**, *54*, 11169–11186. [CrossRef]

38. Perdew, J.P.; Wang, Y. Accurate and simple analytic representation of the electron-gas correlation energy. *Phys. Rev. B* **1992**, *45*, 13244–13249. [CrossRef]

39. Blöchl, P.E. Projector agmented-wave method. *Phys. Rev. B* **1994**, *50*, 17953–17979. [CrossRef]

40. Perdew, J.P.; Burke, K.; Ernzerhof, M. Generalized gradient approximation made simple. *Phys. Rev. Lett.* **1996**, *77*, 3865. [CrossRef]

41. Henkelman, G.; Uberuaga, B.P.; Jonsson, H.A. Climbing image nudged elastic band method for finding saddle points and minimum energy paths. *J. Chem. Phys.* **2000**, *113*, 9901–9904. [CrossRef]

42. Lässer, R.; Bickmann, K. Phase diagram of the Nb-T system. *J. Nucl. Mater.* **1985**, *132*, 244–248. [CrossRef]

43. Baskes, M.I. Modified embedded-atom potentials for cubic materials and impurities. *Phys. Rev. B* **1992**, *46*, 2727. [CrossRef]

44. Strayer, R.W.; Mackie, W.; Swanson, L.W. Work function measurements by the field emission retarding potential method. *Surf. Sci.* **1973**, *34*, 225–248. [CrossRef]

45. Michaelson, H.B. The work function of the elements and its periodicity. *J. Chem. Phys.* **1977**, *48*, 4729–4733. [CrossRef]

46. Gong, L.; Su, Q.; Deng, H.; Xiao, S.; Hu, W. The stability and diffusion properties of foreign impurity atoms on the surface and in the bulk of vanadium: A first-principles study. *Comput. Mater. Sci.* **2014**, *81*, 191–198. [CrossRef]

*applied
sciences*

MDPI

Article

The Electronic, Magnetic, Half-Metallic and Mechanical Properties of the Equiatomic Quaternary Heusler Compounds FeRhCrSi and FePdCrSi: A First-Principles Study

Liefeng Feng [1], Jiannan Ma [1], Yue Yang [1], Tingting Lin [2] and Liying Wang [1,*]

[1] Tianjin Key Laboratory of Low Dimensional Materials Physics and Preparing Technology, Faculty of Science, Tianjin University, Tianjin 300350, China; fengfl@tju.edu.cn (L.F.); majiannan@tju.edu.cn (J.M.); 2018210142@tju.edu.cn (Y.Y.)
[2] Institute of Materials Science, Technische Universtät Darmstadt, 64287 Darmstadt, Germany; ttlin1990@126.com
* Correspondence: liying.wang@tju.edu.cn; Tel.: +86-138-2112-8892

Received: 25 October 2018; Accepted: 16 November 2018; Published: 23 November 2018

Featured Application: Two new 1:1:1:1 quaternary Heusler based half-metals have been designed by means of the first-principles method, and their half-metallic properties are quite robust to the hydrostatic strain or tetragonal distortion.

Abstract: By using the first-principles method, the electronic structures and magnetism of equiatomic quaternary Heusler alloys FeRhCrSi and FePdCrSi were calculated. The results show that both FeRhCrSi and FePdCrSi compounds are ferrimagnets. Both compounds are half-metals and their half-metallicity can be maintained in a wide range of variation of the lattice constant under hydrostatic strain and c/a ratio range under tetragonal distortion, implying that they have low sensitivity to external interference. Furthermore, the total magnetic moments are integers, which are typical characteristics of half-metals. The calculated negative formation energy and cohesive energy indicate that these two alloys have good chemical stability. Furthermore, the value of the elastic constants and the various moduli indicate the mechanical stability of these two alloys. Thus, FeRhCrSi and FePdCrSi are likely to be synthesized in the experiment.

Keywords: first-principles method; half-metallicity; equiatomic quaternary Heusler compounds

1. Introduction

Since the first half-metal (HM) NiMnSb compound was discovered in 1983 [1], the HM materials have attracted more and more scholars and researchers to investigate. The HM material has a very novel electronic structure, showing semiconducting property in on spin channel and metallic property in the other spin channel [2,3]. This special electronic structure results in a perfect 100% spin polarization near the Fermi level, which makes HM a novel spintronic material. Until now, many different structures of materials have been found to be half-metallic, such as Heusler compounds, dilute magnetic semiconductors, and ferromagnetic metallic oxides, etc. [1–8]. Among them, Heusler compounds are of great importance and present many unique physical properties, including the adjustable electronic structure and relative high Curie temperature [9,10], etc.

Heusler compounds, which have a general formula X_2YZ, are a class of intermetallic compounds (also called full-Heusler compounds). In the formula X_2YZ, X and Y atoms are usually transition metals, and Z is a main group element [11]. For full-Heusler compounds, there are two possible, different atomic arrangement fashions: the Cu_2MnAl-type and Hg_2CuTi-type structure [12]. Specially, when X and Y are the same atoms, the $D0_3$ structure is formed [13,14]. And when one of the X atoms is substituted with a vacant, the $C1_b$ structure is formed [15]. When one of the X atoms is substituted by another different transition mental, the equiatomic quaternary Heusler compounds (EQHs) (LiMgPdSn structure) with the chemical formula of XMYZ are formed. Recently, several series of EQHs have been found and attracted wide attention. Compared with the binary and ternary Heusler HMs that have been mentioned above, the biggest advantage of the EQH HMs is their lower dissipation, which is caused by their lesser amount of atomic arrangement disorder in experiment [16].

Felser et al. have predicted several EQHs to be theoretical HM ferromagnets and have synthesized them experimentally, such as CoFeMnZ (Z = Al, Si, Ga, Ge), NiFeMnGa and NiCoMnGa [17–20]. Alijani et al. have showed that NiFeMnGa and NiCoMnGa alloys are HM ferromagnets [14]. Gao et al. Reported that CoFeCrAl and CoFeCrSi compounds are several excellent HM ferromagnets [21]. Karimain et al. found that the NiFeTiP and NiFeTiSi alloys are true HM ferromagnets and that the NiFeTiGe has a near HM character [22]. All the above mentioned EQHs are those with 3d transition elements. Very recently, a series of new EQHs with a 4d transition element have been reported. Some of them are as follows: Berri et al. have investigated the robust half-metallicity of the ZrCoTiZ (Z = Si, Ge Ga and In) [23]. The half-metallic and mechanic characters of ZrRhHfZ (Z = Al, Ga and In) [24] and FeCrRuSi [25] were investigated by Wang et al. Guo et al. have reported the excellent half-metallic property of ZrFeVZ (Z = Al, Ga and In) [26]. The new reports of EQHs with a 4d transition element expand the research scope of the spintronic materials. Thus, we can conclude that 4d-transition-elements-contained HMs are new potential spintronic materials worth researching.

In our present work, the electronic, magnetic, half-metallic and mechanical properties of two newly designed EQH compounds, FeRhCrSi and FePdCrSi, were investigated by using the first-principles method. We also have discussed the effect of the hydrostatic strain and tetragonal distortion on their half-metallic properties. And the calculated results indicate that their half-metalicity can be kept in a wide lattice constant range and c/a ratio range. Moreover, the chemical stability and the mechanical properties of FeRhCrSi and FePdCrSi have also been studied in details.

2. Materials and Methods

The electronic and magnetic structures were calculated by using CASTEP code, which was based on the density functional theory (DFT) [27–29]. The ultrasoft pseudopotential was used to described the interactions between the valence electrons and the atomic core. The exchange-correlation energy was calculated under the generalized gradient approximation (GGA) [30,31]. For all calculated cases, the cut-off energy of 450 eV was chosen. The k-points mesh was set as $10 \times 10 \times 10$ [32,33]. The convergence tolerance for all the calculations was selected as a difference on total energy within 5×10^{-6} eV/atom.

3. Results

3.1. Total Energy and Structural Stability

For the LiMgPdSn-type EQH compounds FeRhCrSi and FePdCrSi, due to the different atomic arrangements of them, there are three possibly different crystal structures (here, named as type-1, type-2, and type-3), shown in Figure 1a–c. The atom positions in Wychoff coordinates are also shown in Table 1. In order to get the ground state properties of FeRhCrSi and FePdCrSi, the geometry optimization was performed by calculating the total energy of the FeRhCrSi and FePdCrSi compounds at different lattice constants for the three possible crystal structures. The calculated total energy (E_t) as the function of lattice constants (a) for FeRhCrSi and FePdCrSi compounds (E_t-a curves) are shown in Figure 2. It can be clearly seen that both FeRhCrSi and FePdCrSi compounds of type-1 structure have the lowest total energy, which means the type-1 structure is the most stable configuration. Considering the different magnetic states of these compounds, the total energy as a function of the lattice constants for the type-1 structure FeRhCrSi and FePdCrSi compounds in ferromagnetic

(FM), ferrimagnetic (FIM), and nonmagnetic (NM) states has been calculated (as shown in Figure 3. From Figure 3, it can be found that both FeRhCrSi and FePdCrSi compounds in the FIM state show the lowest total energy. Therefore, for both FeRhCrSi and FePdCrSi compounds, the type-1 structure in the FIM state was the most stable state. Thus, we will focus on discussing the physical properties of these compounds with a type-1 structure and FIM state in the following parts of this paper. And the obtained equilibrium lattice constants at their ground states for FeRhCrSi and FePdCrSi are 5.82 Å and 5.87 Å, respectively, and the results have been shown in Table 2. The small difference of the optimized lattice constants between FeRhCrSi and FePdCrSi compounds derive from the close ionic radii between Rh and Pd atoms.

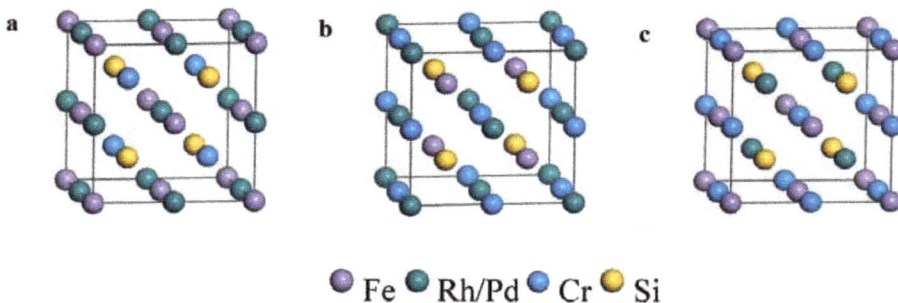

Figure 1. The three possible structures of FeRhCrSi and FePdCrSi (**a**) type-1, (**b**) type-2 and (**c**) type-3.

Table 1. Three possible structures of FeRhCrSi and FePdCrSi.

Structure	Fe	Rh/Pd	Cr	Si
Type 1	4a	4c	4b	4d
Type 2	4b	4c	4a	4d
Type 3	4a	4b	4c	4d

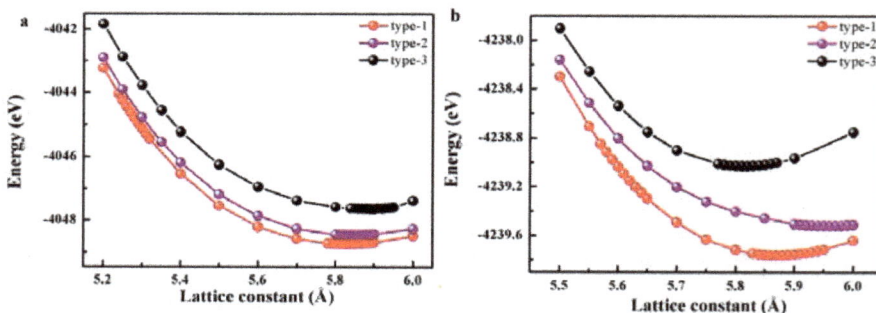

Figure 2. Energy optimization with different crystal structures of (**a**) FeRhCrSi and (**b**) FePdCrSi.

Table 2. The optimized equilibrium lattice constant a_0 (Å), the total and partial magnetic moments (μ_B) at the equilibrium lattice, the formation energy E_f and cohesive energy E_c (eV) for the equiatomic quaternary Heusler (EQH) compounds FeRhCrSi and FePdCrSi.

Compounds	a_0	M_{tot}	M_{Fe}	$M_{Rh/Pd}$	M_{Cr}	M_{Si}	E_f	E_c
FeRhCrSi	5.82	3.00	−0.26	0.22	3.10	−0.06	3.12	−20.41
FePdCrSi	5.87	4.00	0.78	−0.14	3.44	−0.08	−1.97	−18.66

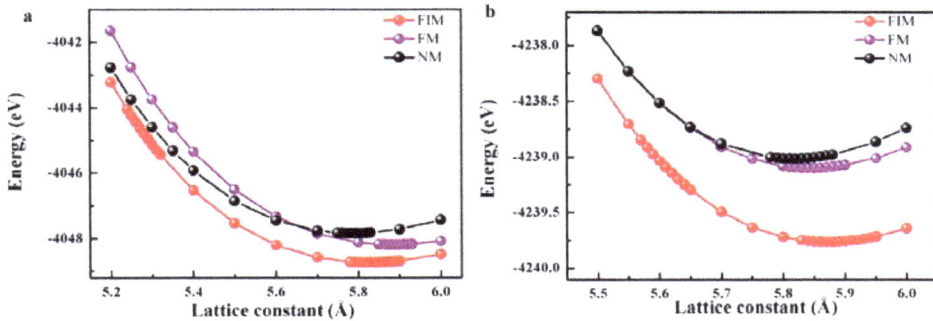

Figure 3. Energy optimization with different magnetic states of (**a**) FeRhCrSi and (**b**) FePdCrSi.

3.2. Electronic, Magnetic, and Half-Metallic Properties

In this part, the electronic and magnetic structure of FeRhCrSi and FePdCrSi at their equilibrium lattice constants have been discussed in detail. The calculated band structures of FeRhCrSi and FePdCrSi compounds have been displayed in Figure 4. From Figure 4a,b, it can be found that both FeRhCrSi and FePdCrSi compounds show quite similar band structure characters. In the majority-spin channel, the Fermi level is located in an indirect band gap. The conduction band minimum (CBM) appears at the X point and the valence band maximum (VBM) occurs at the G point of the Brillouin zone. The values of the indirect band gaps in the majority-spin channel are 0.336 eV and 0.177 eV for EQH compounds FeRhCrSi and FePdCrSi, respectively. However, their bands show a metallic cross over at the Fermi level in the minority-spin channel. This results in their 100% spin polarization. Thus, the EQH compounds FeRhCrSi and FePdCrSi are HMs.

Figure 5a,b show the calculated total and partial density of states (DOS) of FeRhCrSi and FePdCrSi at their equilibrium lattice constants. From Figure 5, one can further understand their electronic structures. It can be seen that there is a metallic overlapping with the Fermi level in the minority-spin channel for both the EQH compounds FeRhCrSi and FePdCrSi. By contrast, in the majority-spin direction, the Fermi level locates in a relative wide energy gap (the half-metallic gap). Therefore, the EQH compounds FeRhCrSi and FePdCrSi are HMs that are consistent with the above discussion of their band structures. For FeRhCrSi and FePdCrSi compounds, because of the similar atomic configuration, they present two quite similar total and partial DOS. Here, we take the FeRhCrSi compound as an example and give a detailed discussion on the DOS. From Figure 5a, it can be found that the DOS is mainly dominanted by the Si atom in the lower energy region (−7 eV–5 eV) for the two spin directions. In the energy region −5 eV to −3 eV, the DOS are mainly occupied by Rh-4d states. In the range of −3 eV to −1 eV, except a small contribution of Rh-4d and Cr-3d states, both spin channels are dominated by Fe-3d states. In the energy region around the Fermi level (−1 eV to 1 eV), the energy peak is the result of the common contribution of Fe, Rh, and Cr atoms with d orbitals. It can be observed that, in the minority spin, the main contribution to the DOS is the strong hybridization among the Fe-3d state, Rh-4d state, and Cr-3d state around the Fermi level. In the majority spin, the energy gap mainly comes from the covalent hybridization between the nearest neighbor atoms: Fe and Cr. Galanakis et al. [34] have reported that the formation of the bonding and antibonding states can arise from the strong covalent hybridization between the lower-energy d states (higher-valent) transition metal atom (such as Fe) and the higher-energy d states (lower-valent) transition metal (such as Cr). And then, the half-metallic band gap generates between the bonding and antibonding bands. It has also been reported that, due to the strong intra-atomic exchange interaction, the low valence transition metal atom usually has a large spin splitting [35]. From Figure 5a, it can be found that, compared with Fe and Rh atoms, the Cr atom with a lower valent has a larger spin splitting. Thus, the Cr atom is the main contributor to the total magnetic moment of the FeRhCrSi compound.

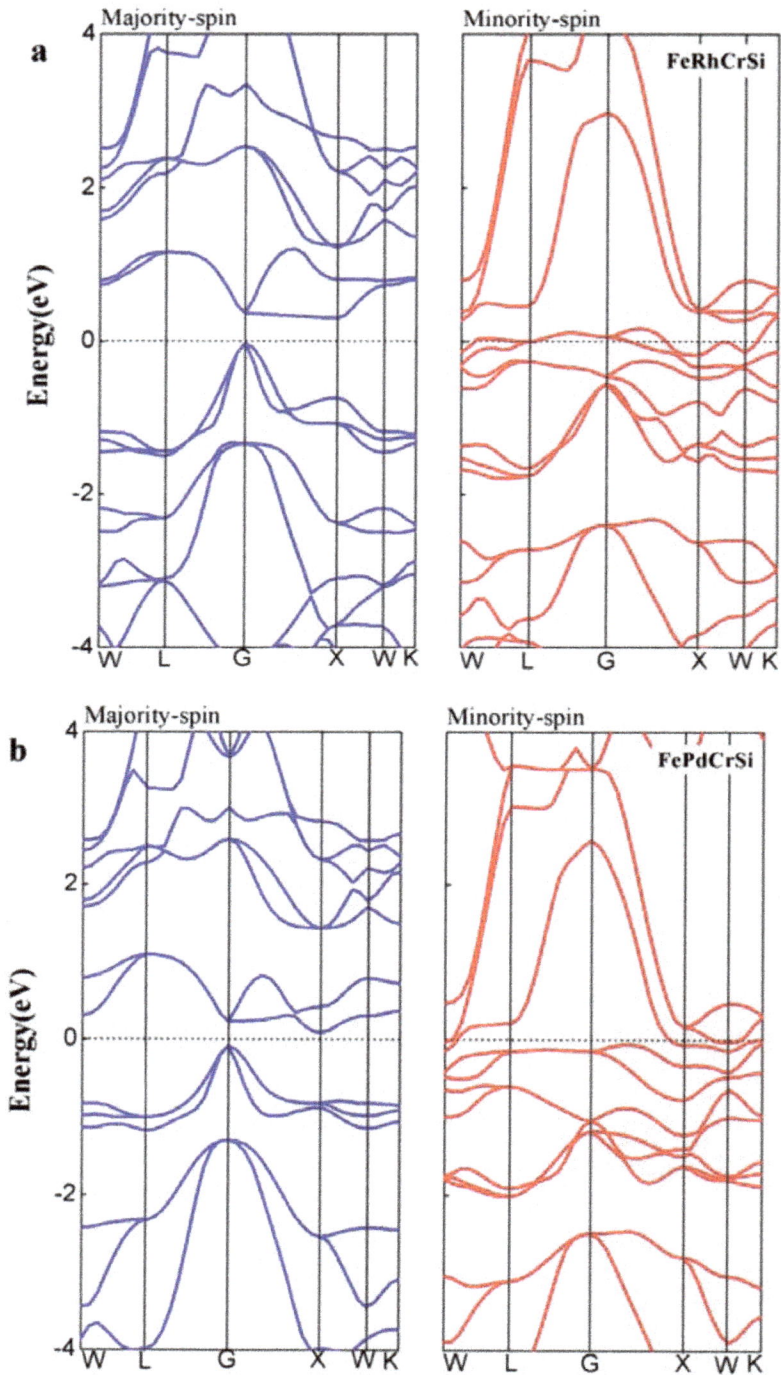

Figure 4. The band structures of (**a**) FeRhCrSi and (**b**) FePdCrSi at their equilibrium lattice constants.

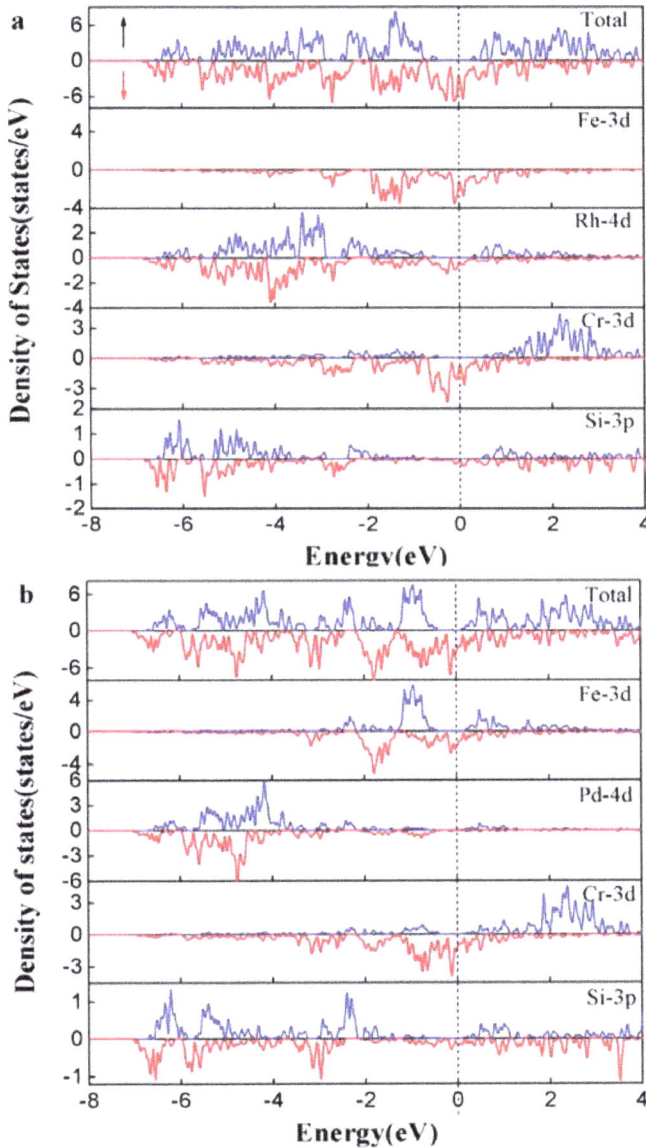

Figure 5. The total and partial density of states of (**a**) FeRhCrSi and (**b**) FePdCrSi at their equilibrium lattice constants.

The total and partial magnetic moments at the equilibrium lattice constant are listed in Table 2. The total magnetic moments (M_t) are integers for both FeRhCrSi and FePdCrSi compounds, which are 3 μB and 4 μB, respectively. The M_t follows the Slater–Pauling behavior, which can be expressed by M_t = Z_t-24, where Z_t is the total number of valence electrons [35,36]. It can be found that the M_t is mainly from the Cr atom (3.10 μB and 3.44 μB for FeRhCrSi and FePdCrSi, respectively) and this is consistent with the above DOS discussion. Meanwhile, Fe, Rh, and Pd atoms have a small contribution to M_t.

Furthermore, due to the anti-parallel atomic magnetic moment among the Fe, Rh/Pd, and Cr atoms, both FeRhCrSi and FePdCrSi compounds exhibit a ferrimagnetic character.

3.3. Structural Stability

In order to test whether FeRhCrSi and FePdCrSi compounds can be synthesized and form a stable phase experimentally, the formation energy E_f and cohesive energy E_c were calculated in this work. The calculated values of the E_f and E_c are listed in Table 2. Taking FeRhCrSi as an example, the formation energy can be expressed by

$$E_f = E_{tot} - E_{Fe}^{bulk} - E_{Rh}^{bulk} - E_{Cr}^{bulk} - E_{Si}^{bulk} \tag{1}$$

Here, E_{tot} is the total energy of FeRhCrSi, E_{Fe}^{bulk}, E_{Rh}^{bulk}, E_{Cr}^{bulk} and E_{Si}^{bulk} are the total energies for each Fe, Rh, Cr, and Si atoms in their bulk states [37]. The bulk states that adopted in our calculation of Fe and Cr are body centered cubic(b.c.c) structure, Pd and Rh are face centered cubic(f.c.c) structure, and Si is diamond cubic(d.c) structure. The cohesive energy is expressed by

$$E_c = E_{tot} - E_{Fe}^{iso} - E_{Rh}^{iso} - E_{Cr}^{iso} - E_{Si}^{iso} \tag{2}$$

Here, E_{Fe}^{iso}, E_{Rh}^{iso}, E_{Cr}^{iso} and E_{Si}^{iso} refer to the energies of each isolated atom [37]. It can be found that formation energies for these alloys are both negative values, which are -3.12 eV for FeRhCrSi and -1.97 eV for FePdCrSi, indicating that these two alloys are stable and cannot be decomposed easily. Furthermore, these alloys have great negative cohesive energies, which are -20.41 eV for FeRhCrSi and -18.66 eV for FePdCrSi, implying that they have good chemical stability in practice and are likely to be synthesized in the experiment.

3.4. Mechanical Properties

In this section, we discuss the mechanical properties of the FeRhCrSi and FePdCrSi compounds. For cubic crystals, there are only three independent single-crystal elastic constants (C_{11}, C_{12} and C_{44}). According to these three elastic constants, other important elastic moduli can be calculated as the following formulas:

$$B = \frac{C_{11} + 2C_{12}}{3} \tag{3}$$

$$G_V = \frac{C_{11} - C_{12} + 3C_{44}}{5} \tag{4}$$

$$G_R = \frac{5C_{44}(C_{11} - C_{12})}{4C_{44} + 3(C_{11} - C_{12})} \tag{5}$$

$$G = \frac{G_V + G_R}{2} \tag{6}$$

$$E = \frac{9GB}{2B + G} \tag{7}$$

$$\delta = \frac{3B - 2G}{2(3B + G)} \tag{8}$$

$$A = \frac{2C_{44}}{C_{11} - C_{12}} \tag{9}$$

Here, B is the bulk modulus, G_V is the Voigt's shear modulus, G_R is the Reuss's shear modulus, G is the shear modulus, E is the Young's modulus, δ is the Possion's ratio, and A is the anisotropy factor.

Based on the above elastic moduli, we can examine the mechanical stability of these two EQHs FeRhCrSi and FePdCrSi according to the following Born and Huang [38] generalized stability criteria:

$$C_{44} > 0; \; \frac{C_{11} - C_{12}}{2}; \; B = \frac{C_{11} + C_{12}}{2} > 0; \; C_{12} < B < C_{11} \tag{10}$$

Table 3 lists all the calculated elastic constants for FeRhCrSi and FePdCrSi. From these calculated values, one can clearly see that our results for C_{11}, C_{12}, and C_{44} follow the generalized stability criteria. Thus, the FeRhCrSi and FePdCrSi compounds are mechanically stable.

Table 3. The calculated elastic constants C_{ij}, bulk modulus B, shear modulus G, Young's modulus E (GPa), Pugh's ratio B/G and anisotropy factor A for the EQH compounds FeRhCrSi and FePdCrSi.

Compounds	C_{11}	C_{12}	C_{44}	B	G	E	B/G	A
FeRhCrSi	294.7	112.9	106.6	173.5	100.0	251.7	1.74	1.17
FePdCrSi	179.4	121.5	75.1	140.8	45.9	124.2	3.07	2.59

Moreover, the Pugh's ratios B/G of FeRhCrSi and FePdCrSi compounds are 1.74 and 3.07, respectively. According to Pugh's criteria, the material tends to be ductile when the Pugh's ratio is larger than 1.75. Otherwise, the material tends to be brittle when the Pugh's ratio is less than 1.75. From Table 3, it can be seen that the Pugh's ratio of FeRhCrSi is less than 1.75, which indicates that FeRhCrSi is brittle. However, the Pugh's ratio of FePdCrSi is larger than 1.75, thus suggesting that FePdCrSi is ductile. Furthermore, the stiffness of the material can be characterized by Young's modulus E. The material will become stiffer when the value of E is higher. Thus, it can be said that the FeRhCrSi is stiffer than FePdCrSi.

3.5. Strain Effect on the Electronic and Magnetic Properties

In fact, the half-metallic character of HMs is very sensitive to external conditions (such as the temperature and train effect) in its application on spintronic devices. The change of the lattice constants can destroy the half-metallicity. Thus, the band structures and magnetism of FeRhCrSi and FePdCrSi under the strain effect (including hydrostatic strain and tetragonal deformation) have been calculated to test the thermal expansion and external strain effect on the half-metallicity of them. Here, the conduction band minimum (CBM) and the valence band maximum (VBM) as functions of the lattice constant and c/a ratio in the majority spin channel were calculated and plotted to describe the change of the half-metallic behavior of the FeRhCrSi and FePdCrSi compounds (as shown in Figures 6 and 7).

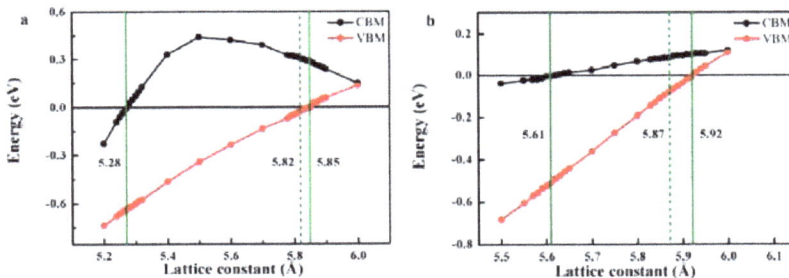

Figure 6. The conduction band minimum (CBM) and valence band maximum (VBM) in the majority spin channel as functions of lattice constants for (**a**) FeRhCrSi and (**b**) FePdCrSi compounds.

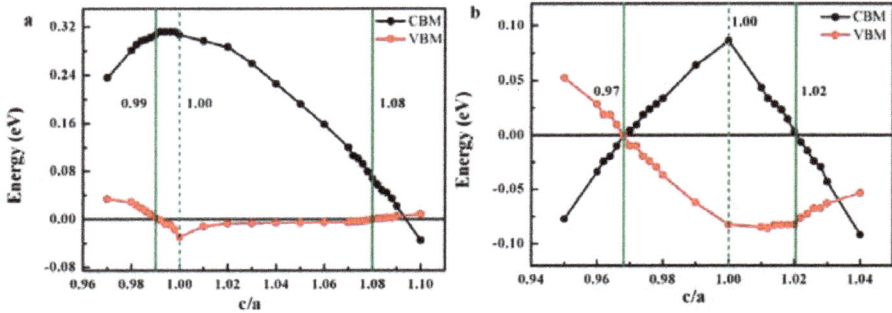

Figure 7. The CBM and VBM in the majority spin channel as functions of c/a ratios for (**a**) FeRhCrSi and (**b**) FePdCrSi compounds.

Figure 6 shows the CBM and VBM values under different hydrostatic strain for FeRhCrSi and FePdCrSi. From Figure 6, it can be obviously found that the half-metallicity of FeRhCrSi and FePdCrSi compounds can be kept in wide lattice constants that range from 5.28 Å–5.85 Å for FeRhCrSi and 5.61 Å–5.92 Å for FePdCrSi, respectively. This indicates that these two EQHs can keep their half-metallic properties in the lattice distortion range of −9.3% to 0.5% and −4.4% to 1% (relative to their equilibrium lattice constants). For FeRhCrSi and FePdCrSi compounds, it can be clearly seen that its half-metallic property is very robust to the lattice constant compression. Their half-metallicity cannot be destroyed until the lattice constant is compressed to 5.28 Å (by −9.3%) for FeRhCrSi and 5.61 Å (by −4.4%) for FePdCrSi, their conduction bands have an overlapping with the Fermi level at 5.28 Å and 5.61 Å. However, the half-metallicity of FeRhCrSi is relatively sensitive to the lattice constant expansion and their half-metallic properties will be destroyed when their lattice constants are expanded by about 0.5% and 1% for FeRhCrSi and FePdCrSi compounds, respectively.

Figure 7a,b show the VBM and CBM values under different c/a ratios in the range of 0.96–1.10 and 0.95–1.04 for FeRhCrSi and FePdCrSi, respectively. In this case, the cubic unit cells are compressed or stretched into a tetragonal one and the volumes of them are kept unchanged. As seen in Figure 7, it seems that their half-metallic properties are also robust to the tetrogonal deformation of lattice constants for FeRhCrSi and FePdCrSi compounds. In detail, FeRhCrSi and FePdCrSi compounds can keep their half-metallic properties when the c/a ratios change from 0.99 to 1.08 and 0.97 to 1.02, respectively. As the c/a ratios of FeRhCrSi and FePdCrSi decreases or increases from 1.0, their CBM value decrease and the value of VBM increases. Finally, this leads to the reduction or disappearance of their half-metallic band gaps. In order to further observe the change of the half-metallic bang gap with the tetragonal distortion, several majority-spin band structures around Fermi level at different c/a ratios for FeRhCrSi compound have been shown in Figure 8. From Figure 8, it can be clearly seen that the half-metallic band gap of FeRhCrSi decreases with tetragonal distortion and disappears at c/a = 0.97 and c/a = 1.02.

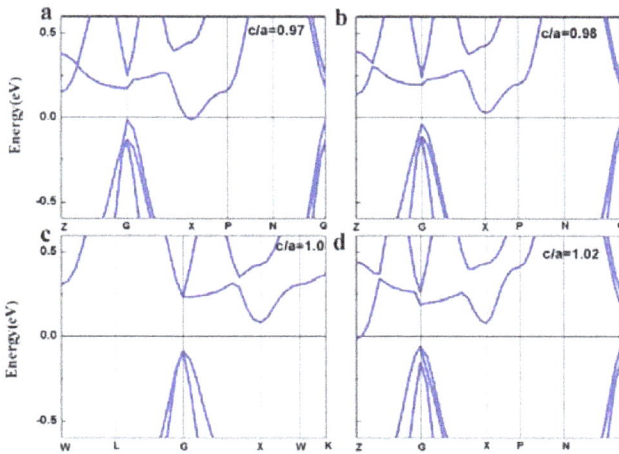

Figure 8. The majority spin band structures around Fermi level at different c/a ratios for FePdCrSi compound (**a**) c/a = 0.97, (**b**) c/a = 0.98, (**c**) c/a = 1.0 and (**d**) c/a = 1.02.

In the next section, we will give a detailed discussion about the effect of the hydrostatic strain and tetragonal deformation on the magnetism of FeRhCrSi and FePdCrSi compounds. In Figures 9 and 10, the curves of the total and partial magnetic moments as functions of lattice constants and c/a ratios for FeRhCrSi and FePdCrSi are given. Clearly, one can observe that the total and partial magnetic moments of FeRhCrSi and FePdCrSi present quite a similar behavior with the change of the lattice constants and c/a ratios. From Figure 9, one can clearly observe that the total magnetic moments of both the FeRhCrSi and FePdCrSi compounds are always fixed into an integer value of 3 μ_B and 4 μ_B in the whole variational range. The atomic moments of the Fe and Cr atoms are very sensitive to the hydrostatic strain. The atomic moment of the Fe atom decreases with the increasing of the lattice constant. Whereas, the atomic moment of the Cr atom shows a growing trend with the increase of the lattice constant. Thus, the total magnetic moments of FeRhCrSi and FePdCrSi compounds can keep a fixed value in the whole variational range of the lattice constant. In Figure 10, one can observe that both the total and atomic magnetic moments of FeRhCrSi and FePdCrSi almost maintain an invariant constant with the variation of c/a ratios. This indicates that the magnetic properties of the FeRhCrSi and FePdCrSi are also not sensitive to the tetraganal deformation.

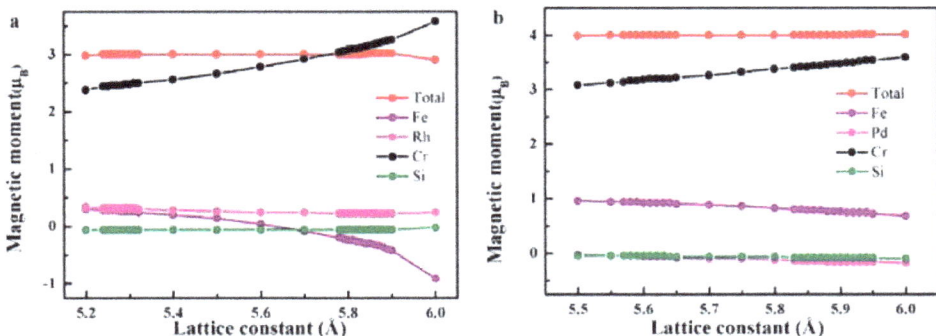

Figure 9. The total and partial magnetic moments as functions of lattice constants for (**a**) FeRhCrSi and (**b**) FePdCrSi.

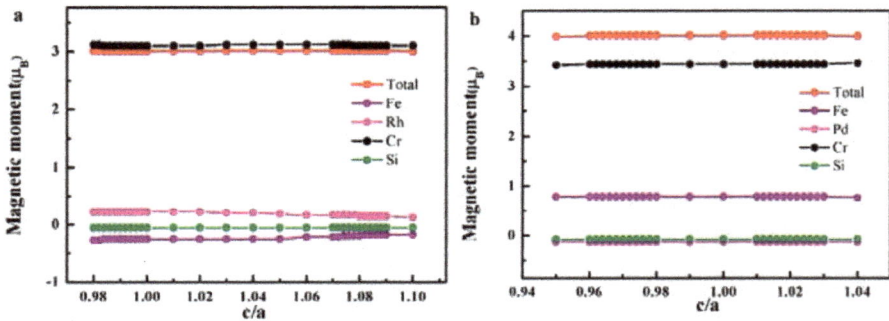

Figure 10. The total and partial magnetic moments as functions of c/a ratios for (**a**) FeRhCrSi and (**b**) FePdCrSi.

4. Conclusions

In summary, we have calculated the electronic, magnetic, half-metallic, structural stability and the mechanical properties of the new designed EQH compounds FeRhCrSi and FePdCrSi by using the first-principles method. The conclusions are listed as below:

(1) FeRhCrSi and FePdCrSi compounds are two new ferrimagnetic HMs with a wide half-metallic band gap of 0.336 eV and 0.177 eV, respectively.

(2) The half-metallicity of the FeRhCrSi and FePdCrSi compounds are very robust to the hydrostatic strain or tetragonal distortion. Especially for FeRhCrSi, the half-metallicity can be kept in a wide lattice constant range (5.28 Å–5.85 Å) under hydrostatic strain and a c/a ratios range (0.98–1.08) under tetragonal distortion, respectively.

(3) The total magnetic moment of FeRhCrSi and FePdCrSi compounds are 3 μ_B and 4 μ_B, respectively, which obey the Slater–Pauling rule: $M_t = Z_t$-24. The main contributor of the total magnetic moments are both Cr atom for FeRhCrSi and FePdCrSi.

(4) The large negative values of the calculated formation energy and cohesion energy show the direct evidence of the chemical and thermal stability for FeRhCrSi and FePdCrSi compounds. This indicates that they are likely to be synthesized in the experiment.

(5) The elastic constants and the various moduli indicate the mechanical stability of FeRhCrSi and FePdCrSi compounds.

Author Contributions: L.Y. Wang designed the project. L.F. Feng and J.N. Ma performed the calculations and prepared the manuscript. All authors analyzed the data and discussed the results.

Funding: This work was funded by Natural Science Foundation of Tianjin City (No.16JCYBJC17200, 17JCQNJC02800), and National Nature Science Foundations of China (No. 51701138).

Conflicts of Interest: The authors declare no conflicts of interest.

References

1. De Groot, R.A.; Mueller, F.M.; van Engen, P.G.; Buschow, K.H.J. New class of materials: half-metallic ferromagnets. *Phys. Rev. Lett.* **1983**, *50*, 2024. [CrossRef]
2. Wang, X.; Cheng, Z.; Jin, Y.; Wu, Y.; Dai, X.; Liu, G. Magneto-electronic properties and tetragonal deformation of rare-earth-element-based quaternary Heusler half-metals: A first-principles prediction. *J. Alloys Compd.* **2018**, *734*, 329–341. [CrossRef]
3. Sun, Q.; Kioussis, N. Prediction of manganese trihalides as two-dimensional Dirac half-metals. *Phys. Rev. B* **2018**, *97*, 094408. [CrossRef]
4. Davatolhagh, S.; Dehghan, A. Dirac-like half-metallicity of $d^0 − d$ half-Heusler alloys. *Phys. C Supercond. Appl.* **2018**, *552*, 53–56. [CrossRef]

5.   Han, Y.; Wang, X. First-Principles Investigation of Half-Metallic Ferromagnetism of a New 1:1:1:1 Type Quaternary Heusler Compound YRhTiSi. *J. Supercond. Novel Magn.* **2018**. [CrossRef]

6.   Zhang, L.; Wang, X.T.; Rozale, H.; Gao, Y.C.; Wang, L.Y.; Chen, X.B. Electronic structures, magnetic properties and half-metallicity in Heusler alloys Zr_2IrZ (Zá = áAl, Ga, In). *Curr. Appl. Phys.* **2015**, *15*, 1117–1123. [CrossRef]

7.   Yu, W.; Zhu, Z.; Niu, C.Y.; Li, C.; Cho, J.H.; Jia, Y. Dilute Magnetic Semiconductor and Half-Metal Behaviors in 3 d Transition-Metal Doped Black and Blue Phosphorenes: A First-Principles Study. Nanoscale research letters. *Nanoscale Res. Lett.* **2016**, *11*, 77. [CrossRef] [PubMed]

8.   Galanakis, I.; Mavropoulos, P. Zinc-blende compounds of transition elements with N, P, As, Sb, S, Se, and Te as half-metallic systems. *Phys. Rev. B* **2003**, *67*, 104417. [CrossRef]

9.   Wurmehl, S.; Fecher, G.H.; Kandpal, H.C.; Ksenofontov, V.; Felser, C.; Lin, H.J. Investigation of Co 2 Fe Si: The Heusler compound with highest Curie temperature and magnetic moment. *Appl. Phys. Lett.* **2005**, *88*, 032503. [CrossRef]

10.   Kanomata, T.; Kyuji, S.; Nashima, O.; Ono, F.; Kaneko, T.; Endo, S. The Curie temperature in Heusler alloys Ni_2MnZ (Z = Ga, Sn and Sb) under high pressure. *J. Alloys Compd.* **2012**, *518*, 19–21. [CrossRef]

11.   Campbell, C.C.M. Hyperfine field systematics in Heusler alloys. *J. Phys. F: Met. Phys.* **1975**, *5*, 1931. [CrossRef]

12.   Peng, H.; Perdew, J.P. Rehabilitation of the Perdew-Burke-Ernzerhof generalized gradient approximation for layered materials. *Phys. Rev. B* **2017**, *95*, 081105. [CrossRef]

13.   Wang, R.; Zhao, Y.; Li, Z.; Chen, H.; Tao, X.; Ouyang, Y. The effect of Al content on the structural, mechanical, and thermal properties of B_2-FeAl and $D0_3$-Fe$_3$Al from atomistic study. *Mater. Res. Express* **2018**, *5*, 026512. [CrossRef]

14.   Jamer, M.E.; Wang, Y.J.; Stephen, G.M.; McDonald, I.J.; Grutter, A.J.; Sterbinsky, G.E.; Arena, D.A.; Borchers, J.A.; Kirby, B.J.; Lewis, L.H.; et al. Compensated Ferrimagnetism in the Zero-Moment Heusler Alloy Mn 3 Al. *Phys. Rev. Appl.* **2017**, *7*, 064036. [CrossRef]

15.   Helmholdt, R.B.; de Groot, R.A.; Mueller, F.M.; van Engen, P.G.; Buschow, K.H.J. Magnetic and crystallographic properties of several C1b type Heusler compounds. *J. Magn. Magn. Mater.* **1984**, *43*, 249–255. [CrossRef]

16.   Bainsla, L.; Suresh, K.G. Equiatomic quaternary Heusler alloys: A material perspective for spintronic applications. *Appl. Phys. Rev.* **2016**, *3*, 031101. [CrossRef]

17.   Dai, X.; Liu, G.; Fecher, G.H.; Felser, C.; Li, Y.; Liu, H. New quarternary half metallic material CoFeMnSi. *J. Appl. Phys.* **2009**, *105*, 07E901. [CrossRef]

18.   Klaer, P.; Balke, B.; Alijani, V.; Winterlik, J.; Fecher, G.H.; Felser, C.; Elmers, H.J. Element-specific magnetic moments and spin-resolved density of states in CoFeMn Z (Z = Al, Ga; Si, Ge). *Phys. Rev. B* **2011**, *84*, 144413. [CrossRef]

19.   Alijani, V.; Ouardi, S.; Fecher, G.H.; Winterlik, J.; Naghavi, S.S.; Kozina, X.; Stryganyuk, G.; Felser, C.; Ikenaga, E.; Yamashita, Y.; et al. Electronic, structural, and magnetic properties of the half-metallic ferromagnetic quaternary Heusler compounds CoFeMn Z (Z = Al, Ga, Si, Ge). *Phys. Rev. B* **2011**, *84*, 224416. [CrossRef]

20.   Alijani, V.; Winterlik, J.; Fecher, G.H.; Naghavi, S.S.; Felser, C. Quaternary half-metallic Heusler ferromagnets for spintronics applications. *Phys. Rev. B* **2011**, *83*, 184428. [CrossRef]

21.   Gao, G.Y.; Hu, L.; Yao, K.L.; Luo, B.; Liu, N. Large half-metallic gaps in the quaternary Heusler alloys CoFeCrZ (Z = Al, Si, Ga, Ge): A first-principles study. *J. Alloys Compd.* **2013**, *551*, 539–543. [CrossRef]

22.   Karimian, N.; Ahmadian, F. Electronic structure and half-metallicity of new quaternary Heusler alloys NiFeTiZ (Z = Si, P, Ge, and As). *Solid State Commun.* **2015**, *223*, 60–66. [CrossRef]

23.   Berri, S.; Ibrir, M.; Maouche, D.; Attallah, M. Robust half-metallic ferromagnet of quaternary Heusler compounds ZrCoTiZ (Z = Si, Ge, Ga and Al). *Comput. Condens. Matter* **2014**, *1*, 26–31. [CrossRef]

24.   Wang, X.T.; Cheng, Z.X.; Guo, R.K.; Wang, J.L.; Rozale, H.; Wang, L.Y.; Yu, Z.Y.; Liu, G.D. First-principles study of new quaternary Heusler compounds without 3d transition metal elements: ZrRhHfZ (Z= Al, Ga, In). *Mater. Chem. Phys.* **2017**, *193*, 99–108. [CrossRef]

25. Wang, X.T.; Khachai, H.; Khenata, R.; Yuan, H.K.; Wang, L.Y.; Wang, W.H.; Bouhemadou, A.; Hao, L.Y.; Dai, X.F.; Guo, R.K.; et al. Structural, electronic, magnetic, half-metallic, mechanical, and thermodynamic properties of the quaternary Heusler compound FeCrRuSi: A first-principles study. *Sci. Rep.* **2017**. [CrossRef] [PubMed]

26. Guo, R.; Liu, G.; Wang, X.; Rozale, H.; Wang, L.; Khenata, R.; Wu, Z.; Dai, X. First-principles study on quaternary Heusler compounds ZrFeVZ (Z = Al, Ga, In) with large spin-flip gap. *RSC Adv.* **2016**, *6*, 109394–109400. [CrossRef]

27. Segall, M.D.; Lindan, P.J.D.; Probert, M.J.; Pickard, C.J.; Hasnip, P.J. First-principles simulation: ideas, illustrations and the CASTEP code. *J. Phys. Condens. Matter* **2002**, *14*, 2717. [CrossRef]

28. Clark, S.J.; Segall, M.D.; Pickard, C.J.; Hasnip, P.J.; Probert, M.I.J.; Refson, K.; Payne, M.C. First principles methods using CASTEP. *Zeitschrift für Kristallographie-Cryst. Mater.* **2005**, *220*, 567–570. [CrossRef]

29. Cheeseman, J.R.; Frisch, M.J.; Devlin, F.J.; Stephens, P.J. Hartree-Fock and density functional theory ab initio calculation of optical rotation using GIAOs: Basis set dependence. *J. Phys. Chem. A* **2000**, *104*, 1039. [CrossRef]

30. Perdew, J.P.; Chevary, J.A.; Vosko, S.H.; Jackson, K.A.; Pederson, M.R.; Singh, D.J.; Fiolhais, C. Atoms, molecules, solids, and surfaces: Applications of the generalized gradient approximation for exchange and correlation. *Phys. Rev. B* **1992**, *46*, 6671. [CrossRef]

31. Steinmann, S.N.; Csonka, G.; Corminboeuf, C. Unified Inter- and Intramolecular Dispersion Correction Formula for Generalized Gradient Approximation Density Functional Theory. *J. Chem. Theory Comput.* **2009**, *5*, 2950. [CrossRef] [PubMed]

32. Umari, P.; Pasquarello, A. Polarizability and dielectric constant in density-functional supercell calculations with discrete k-point samplings. *Phys. Rev. B* **2003**, *68*, 085114. [CrossRef]

33. Rath, J.; Freeman, A.J. Generalized magnetic susceptibilities in metals: Application of the analytic tetrahedron linear energy method to Sc. *Phys. Rev. B* **1975**, *11*, 2109. [CrossRef]

34. Galanakis, I.; Dederichs, P.H.; Papanikolaou, N. Slater-Pauling behavior and origin of the half-metallicity of the full-Heusler alloys. *Phys. Rev. B* **2002**, *66*, 174429. [CrossRef]

35. Galanakis, I. Slater-Pauling Behavior in Half-Metallic Magnets. *J. Surf. Interface. Mater.* **2014**, *2*, 74–78. [CrossRef]

36. Fecher, G.H.; Kandpal, H.C.; Wurmehl, S.; Felser, C.; Schönhense, G. Slater-Pauling rule and Curie temperature of Co2Co2-based Heusler compounds. *J. Appl. Phys.* **2006**, *99*, 08J106. [CrossRef]

37. Srivastava, G.P.; Weaire, D. The theory of the cohesive energies of solids. *Adv. Phys.* **1987**, *36*, 463. [CrossRef]

38. Huang, K.; Born, M. Clarendon. In *Dynamical Theory of Crystal Lattices*; Clarendon Press: Oxford, UK, 1954; p. 420.

applied
sciences

MDPI

Article

Structure, Magnetism, and Electronic Properties of Inverse Heusler Alloy Ti₂CoAl/MgO(100) Herterojuction: The Role of Interfaces

Bo Wu [1,2,*], **Haishen Huang [1,3]**, **Guangdong Zhou [4]**, **Yu Feng [5]**, **Ying Chen [6]** and **Xiangjian Wang [7]**

[1] School of Physics and Electronic Science, Zunyi Normal University, Zunyi 563002, China; haishenh@yeah.net
[2] School of Marine Science and Technology, Northwestern Polytechnical University, Xi'an 710072, China
[3] School of Physics, Beijing Institute of Technology, Beijing 100081, China
[4] Guizhou Institute of Technology, Guiyang 550003, China; gdzhou132@163.com
[5] School of Physics and Electronic Engineering, Jiangsu Normal University, Xuzhou 221116, China; fengyu9519@163.com
[6] School of Mathematics and Physics, Anshun University, Anshun 561000, China; ychenjz@163.com
[7] Applied Physics, Division of Materials Science, Department of Engineering Sciences and Mathematics, Luleå University of Technology, SE-971 87 Luleå, Sweden; xiangjian.wang@ltu.se
* Correspondence: fqwubo@163.com; Tel.: +86-0851-2892-7153

Received: 26 October 2018; Accepted: 19 November 2018; Published: 22 November 2018

Abstract: In this study, the interface structures, atom-resolved magnetism, density of states, and spin polarization of 10 possible atomic terminations in the Ti₂CoAl/MgO(100) heterojunction were comprehensively investigated using first-principle calculations. In the equilibrium interface structures, the length of the alloy–Mg bond was found to be much longer than that of the alloy–O bond because of the forceful repulsion interactions between the Heusler interface atoms and Mg atoms. The competition among d-electronic hybridization, d-electronic localization, and the moving effect of the interface metal atoms played an important role in the interface atomic magnetic moment. Unexpected interface states appeared in the half-metallic gap for all terminations. The "ideal" half-metallicity observed in the bulk had been destroyed. In TiAl–Mg and AlAl–O terminations, the maximal spin polarization of about 65% could be reserved. The tunnel magnetoresistance (TMR) value was deduced to be lower than 150% in the Ti₂CoAl/MgO(100) heterojunction at low temperature.

Keywords: Heusler alloy; interface structure; magnetism; spin polarization

---

1. Introduction

The magnetic tunnel junction (MTJ) usually has a large tunnel magnetoresistance (TMR) value and has become a key component of many advanced magnetic devices, such as the read heads in hard-disk drives [1–4], magnetoresistive random access memories [5], and the next generation of high-density, nonvolatile memories and logic devices [6–10] The core component of MTJ is a "sandwich" stack usually consisting of two ferromagnetic layers and a seeding layer. The direction of the two magnetizations of the ferromagnetic layers can be switched individually by an external magnetic field. If the magnetizations are in a parallel orientation, the electrons tend to tunnel through the seeding layer more than if they are in the antiparallel orientation. Consequently, such a MTJ can be switched between two states of electrical resistance—one with low and one with very high resistance. In MTJ, the TMR value is the most important parameter, which can be defined by the Julliere formula with the spin polarizations of the ferromagnetic electrodes as follows [11]:

$$\text{TMR} = 2P_1 P_2 / (1 - P_1 P_2), \tag{1}$$

where P_1 and P_2 denote the spin-polarized values of two ferromagnetic layers. The P value can be defined as follows:

$$P = (N_\uparrow - N_\downarrow)/(N_\uparrow + N_\downarrow), \tag{2}$$

where $N_\uparrow$ and $N_\downarrow$ are the spin-up and spin-down state densities at the Fermi level, respectively. Obviously, the key technologies for MTJ applications aim to increase the spin polarization of ferromagnetic films in MTJ devices. However, achieving such aim is a very complicated task. Atomic disordering [12,13], surface states [14], interface states [15], and thermal effects [16–18] are the fatal reasons for depolarizing ferromagnetic films.

As a major breakthrough in this field, the fabrication of epitaxial Heusler MTJ can exploit coherent electronic tunneling to produce a large TMR, even at room temperature. In 2004, Parkin and co-workers obtained TMR values of up to 220% at room temperature and 300% at low temperatures for CoFeB/MgO/CoFeB MTJ-oriented (100) MgO tunnel barriers and CoFe electrodes [19]. In 2011, a TMR value of about 600% was reported for $Co_2MnAlSi/MgO/CoFe$ MTJ [20]. The current record for the highest Heusler TMR observed experimentally is held by $Co_2MnSi/MgO/Co_2MnSi$, which produced a TMR ratio of 1995% at 4 K [21]. From the ab initio calculation, a theoretical TMR value of about 10^6 was detected in $Co_2CrSi/Cu_2CrAl/Co_2CrSi$ [22]. Recently, an extremely high TMR value exceeding 25,000% was detected in a binary all-Heusler stack $Fe_3Al/BiF_3/Fe_3Al$ at low bias [23]. Therefore, the MgO-based Heusler MTJ holds great potential for TMR device applications.

Heusler alloys are a large family of binary (X_3Z), ternary (X_2YZ), and quaternary (X_1X_2YZ) compounds with more than 1500 known members and an impressive range of properties [24]. In these compounds, X and Y generally denote a transition metal, while Z denotes an sp-element or a main group element. In general, ternary Heusler alloys have two high-ordered structures, namely, the classic Cu_2MnAl-type structure and the newly discovered Hg_2CuTi-type structure. The conventional Cu_2MnAl-type Heusler alloy X_2YZ with FM-3M space group consists of four fcc sublattices. In Wyckoff coordinates, X is located at (0, 0, 0) and (0.5, 0.5, 0.5), Y is located at (0.25, 0.25, 0.25), and Z is located at (0.75, 0.75, 0.75). Each X atom has four Y and Z atoms as nearest neighbors, thereby presenting the same X atomic surrounding. Unlike the Cu_2MnAl-type structure, the Hg_2CuTi-type Heusler alloy, also known as the inverse Heusler alloy, has composition X_2YZ bear F-43M space group. The neighbor X atoms occupy the locations at (0, 0, 0) and (0.25, 0.25, 0.25), the residual Y enters (0.5, 0.5, 0.5), and Z is located at (0.75, 0.75, 0.75). Given the different surroundings of neighboring X atoms, the d-states hybridization must be remodulated according to magnetic atomic electronegativity. Such remodulation results in novel magnetism and electronic properties and surface or interface behaviors.

Given that high Curie temperature and magnetism follow the Slater–Pauling rule, the Cu_2MnAl-type Heusler alloys have received attention in earlier studies. Many Cu_2MnAl-type Heusler alloys, such as Co_2MnX (X = Si, Ge, Sn) [25], Mn_2CoZ (Z = Al, Ga, In, Si, Ge, Sn, Sb), [26] and Cr_2MnZ (Z = P, As, Sb, and Bi) [27], have been predicted with half-metallicity from theoretical ab initio calculations and experimental investigations. TMR experiments of the classic Heusler compound MTJs highlighted significant magnetoresistive effects of these junctions and their potential applications [21,22,28]. However, inverse Ti-based Heusler alloys, such as Ti_2VAl [29], Ti_2FeIn [30], and Ti_2MnGa [31], have been previously predicted with "ideal" half-metallicity in the bulk phase from first-principle calculations. The relatively high experimental TMR values of approximately 262% at a low temperature and 159% at room temperature have been reported for the Fe-based inverse Heusler $Fe_2CoSi/MgO/Co_3Fe$ [32]. However, research on the MTJs of inverse Heusler alloys is still in its infancy. Moreover, to the best of our knowledge, the theoretical or experimental evidence supporting high spin polarization in MTJs for inverse Heusler alloy systems remains scarce.

In our previous work, the inverse Heusler alloy Ti_2CoAl was intensively studied owing to its theoretical 100% spin polarization in the bulk phase from first-principle calculations. The doping effects of d-electrons on the electronic structure and magnetism of the inverse Heusler alloy Ti_2CoAl were also investigated by substituting Nb and V atoms with Ti(A) and Ti(B) atoms. The doped compounds $Ti_{1.25}V_{0.75}CoAl$ and $Ti_{1.5}Nb_{0.5}CoAl$ effectively inhibited the spin-flip excitation and were shown to be promising candidates for spintronic applications [33]. We also investigated the effect of swap, antisite, and vacancy defects on the inverse Heusler alloy Ti$_2$CoAl. A Ti vacancy and a high spin polarization of around 95% were observed in the Co–Al swap [12]. For Heusler MTJ applications, evidence of high spin polarization on the surface are crucial because the potential surface states appearing in the minority spin gap can easily destruct the "ideal" spin polarization in the bulk phase, as has been reported many times in Cu$_2$MnAl-type Heusler alloys. Moreover, (100) surfaces have been comprehensively detected for the Ti$_2$CoAl system by researchers. Given the surface states, the calculated surfaces failed to preserve the half-metallicity observed in the bulk, and high surface spin polarizations were predicted in only the CoCo and AlAl terminations [34].

To obtain direct evidence of inverse Heusler MTJs, we extended our research to Ti$_2$CoAl heterojunctions. Given its popular use as a binary semiconductor and its well-matching structure with that of Heusler alloys, MgO was selected as the seeding layer in Ti$_2$CoAl MTJ. Therefore, the Ti$_2$CoAl(100)/MgO interfaces were further examined in this work. Given the fact that interface atomic disorder can significantly change the spin polarization of Heusler MTJs [14], apart from the standard epitaxial Heusler terminated surfaces cleaved along the Miller indices (100) crystal direction (i.e., TiCo and TiAl terminations), the modified artificial terminations that cape the pure atoms (i.e., TiTi, CoCo, and AlAl terminations) were also examined to extensively search for possible films with a high spin polarization. In this work, inverse Heusler surfaces were epitaxially grown on an MgO(100) substrate to create possible Ti$_2$CoAl(100)/MgO heterojunctions. To understand the physical and chemical properties of the inverse alloy Ti$_2$CoAl/MgO interface, the structures, magnetism, and electronic properties of Ti$_2$CoAl(100)/MgO heterojunctions with varying atomic interfaces were comprehensively investigated.

2. Structures and Calculation Methods

In the calculations, the Ti$_2$CoAl bulks with an Hg$_2$CuTi structure were geometrically optimized to find the minimal energy structures. Afterward, the optimized bulk structure was cleaved along the Miller indices (100) crystal direction to create all cases of "ideal" epitaxial terminated surfaces, namely, TiCo and TiAl terminations. The modified TiTi, CoCo, and AlAl terminations were created by the surface atoms that act as substitutes for the Ti, Co, and Al atoms in the "ideal" terminations. In the interface calculations, nine and seven atomic layers were taken for Heusler alloys and MgO, respectively, and they were connected with each other to form a supercell. When all possible atomic interfaces came in contact with one another, 10 potential Ti$_2$CoAl/MgO(100) junctions were created, as shown in Figure 1. The thicknesses of these junctions were large enough for the central regions. The tested calculations revealed that the atomic moments in the middle layer were extremely close to the bulk values. In the interface calculations, we only focused on the region of three interface layers in the heterojunction given that they produced the greatest influence on the electronic and magnetic properties of Ti$_2$CoAl/MgO(100) junctions.

All calculations were performed using the CASTEP Package and by adopting the density functional theory (DFT). The exchange–correlation interaction was described by performing Perdew–Burke–Ernzerhof (PBE) generalized gradient approximation (GGA) [35]. To deal with the electron–ion interaction, we adopted Vanderbilt-type ultrasoft pseudopotentials [36] and the valence electron configurations of Ti ($3d^2 4s^2$), Co ($3d^7 s^2$), Al ($3s^2 3p^1$), Mg ($3s^2$), and O ($2s^2 2p^4$). For the optimizations of the bulks, we initially assumed that all alloys were ferromagnetic and then applied spin polarization and the $7 \times 7 \times 7$ mesh of special k-points in the Brillouin zone. In the self-consistent calculation, we selected the refined 1×10^{-6} eV/atom and 360 eV as the self-consistent field (SCF) convergence criterion and energy cutoff, respectively. When the positions of atoms were relaxed, we set a convergence criterion of 0.02 eV/Å. To investigate the electronic density of states, we inserted a refined 0.3 Å grid space between k-points in the Brillouin zone. For the interface calculations, we geometrically optimized all supercells using the same parameters employed in the bulk calculation. All technical parameters were tested carefully to ensure the accuracy of the results.

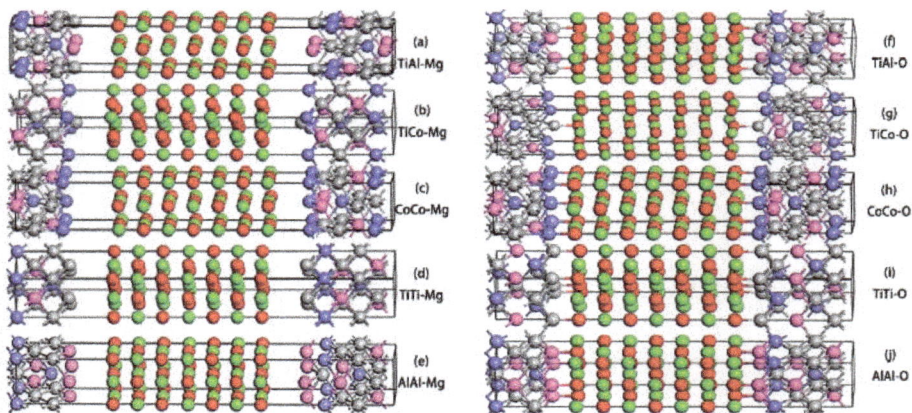

Figure 1. 10 lowest energy terminations in the Ti$_2$CoAl/MgO(100) heterojunction after geometry optimization. (**a**) TiAl–Mg termination; (**b**) TiCo–Mg termination; (**c**) CoCo–Mg termination; (**d**) TiTi–Mg termination; (**e**) AlAl–Mg termination; (**f**) TiAl–O termination; (**g**) TiCo–O termination; (**h**) CoCo–O termination; (**i**) TiTi–O termination; and (**j**) AlAl–O termination. Color code: O (red), Mg (green), Ti (gray), Co (blue) and Al (pink).

3. Results and Discussion

3.1. Interface Structures

The structures in Figure 1 represent the lowest-energy configurations after the geometry optimization. The interface atomic layers were planar before the optimization, and the optimized atomic layers were uneven due to the different atomic interactions. As can be seen in Figure 1a–e, in the alloy–Mg terminations, all interface atoms, especially the Co atoms, demonstrated an inward movement, which was similar to their surface behaviors in Heusler alloys [14,34]. In the Heusler terminations where the interface atoms came in contact with the top Mg atoms, the Heusler and MgO atomic layers were repelled very far. For the alloy–O terminations, the distance between the Heusler and MgO layers was reduced due to the atomic bonding interaction at the interface layers (see Figure 1f–j). The shrinking phenomenon of interface atoms could still be observed.

Table 1 lists the bond types and lengths at different interfaces. The Co–Mg bond in the CoCo–Mg termination had the longest length, while the Al–O bond in the AlAl–O termination had the shortest length. The bond lengths in the alloy–Mg terminations were obviously longer than those in the alloy–O terminations because of the forceful repulsion interactions between the Heusler interface atoms and the Mg atoms, especially those between the metal and Mg atoms. Moreover, the interface repulsion interactions of metal–Mg or metal–O were more vigorous than those of Al–Mg or Al–O. In the TiCo–Mg and TiCo–O terminations, the bond lengths of Co–Mg and Co–O were longer than those of Ti–Mg and Ti–O, thereby creating an uneven interface atomic layer that led to further spin electronic scattering from the mechanical mismatch interface layers to weaken the spin polarization [37].

Table 1. Bond lengths at the interfaces.

Interface Termination	Bond Type	Bond Length (Å)
CoCo–Mg	Co–Mg	4.28
TiTi–Mg	Ti–Mg	4.05
AlAl–Mg	Al–Mg	3.99
TiAl–Mg	Al–Mg	3.99
	Ti–Mg	4.06
TiCo–Mg	Al–Mg	3.55
	Co–Mg	4.02
CoCo–O	Co–O	2.07
TiTi–O	Ti–O	2.14
AlAl–O	Al–O	2.06
TiAl–O	Al–O	2.16
	Ti–O	2.15
TiCo–O	Ti–O	2.08
	Co–O	2.83

3.2. Interface Magnetic Behaviors

Table 2 summarizes the atom-resolved magnetic moments (AMMs) at the interface and subinterface layers in Ti_2CoAl and at the interface layer in MgO for various terminations of the Ti_2CoAl/MgO(100) interface. To facilitate comparisons with the bulk and surface values, the calculated atom-resolved magnetic moments per cell in Ti_2CoAl bulk and (100) surfaces (TiAl termination and TiCo termination) are also listed in this table. The AMMs from the middle layers were close to the corresponding bulk values, thereby indicating that the implemented GGA + PBE scheme could reliably deal with the Heusler hertrojunction system. The middle layer is also called the bulk-like layer. Following the reduction of atomic coordination numbers at the surfaces, the crystal field was weakened, and the localization of d-electron atoms were enhanced, thereby resulting in the rehybridization of Ti and Co atoms. As a result, the surface Ti AMM was obviously enhanced when compared with the bulk value. For the surface and subsurface Co atoms, the AMM slightly decreased owing to the enhanced d-electronic hybridization caused by the surface Co atomic shrink. This same result had also been obtained in our previous work [34]. Similar surface behaviors were also observed in $Ti_2FeGe(001)$ [38] and $Co_2MnGe(111)$ [39].

Interface AMMs are various and complex. The Ti may be located at the (0, 0, 0) or (0.25, 0.25, 0.25) sites and presents different AMMs in the bulk [29]. Table 2 shows that in the alloy–Mg terminations, the interface and subinterface Ti AMMs were slightly larger than those at the middle layers, thereby suggesting that the d-electron localization originating from relatively large metal–Mg bond lengths was enhanced. However, for the interface Co atom, the fierce inward shrink promoted the d-electron hybridization to resist d-electronic localization. Therefore, the interface Co AMMs, except for the CoCo–Mg termination, slightly decreased. In the alloy–O terminations, the relatively short metal–O bond lengths reduced the part of d-electron localization, thereby leading to a remarkable direct magnetic exchange. Therefore, the interface or subinterface Ti and Co AMMs in the alloy–O terminations, except for the CoCo–O termination, had low values. The modified CoCo–O or CoCo–Mg terminations were created by the surface atoms substituted for the Ti atoms in the "ideal" TiCo–Mg or TiCo–O terminations. Therefore, the two interface Co atoms had different magnetic properties. In the CoCo–Mg and CoCo–O terminations, given that the periodic structure of the Heusler crystal field was cut off and considering the competition between the localization and hybridization of d-electrons, the interface Co AMM was a large value, especially in the CoCo–Mg termination. Meanwhile, for the interface Al atom, the absolute value of AMMs decreased as a result of the reduced magnetic atoms. In MgO films, the interface or subinterface Co and Mg atoms suffered from an extremely small spin polarization and had zero spin magnetism in most cases.

Table 2. Atom-resolved magnetic moments at interface and subinterface layers in Ti$_2$CoAl and the interface layer in MgO. The number following "*" denotes the atoms at the subinterface, while the number enclosed in brackets denotes the magnetism coming from different atomic sites.

Termination	Layers	Ti	Co	Al	O	Mg
bulk		0.94 (1.68)	−0.48	−0.14	0.00	0.00
TiAl	(100) surface	1.08	−0.36	−0.18	0.00	0.00
TiCo	(100) surface	1.70	−0.20	−0.10	0.00	0.00
TiAl–Mg	interface	1.10	−0.40 *	−0.20	0.04	0.00
	middle layer	0.88 (1.59)	−0.32	−0.12	0.00	0.00
TiCo–Mg	interface	1.64	−0.07	0.09 *	0.02	0.02
	middle layer	0.82 (1.62)	−0.24	−0.12	0.00	0.00
CoCo–Mg	interface	−0.20 *	1.26(0.16)	−0.06 *	0.00	0.00
	middle layer	0.84 (1.66)	−0.34	−0.12	0.00	0.00
TiTi–Mg	interface	1.02	0.28 *	−	0.02	0.03
	middle layer	0.86 (1.66)	−0.34	−0.14	0.00	0.00
AlAl–Mg	interface	0.34 *	0.40 *	−0.02	0.00	0.00
	middle layer	0.86 (1.6)	−0.34	−0.12	0.00	0.00
TiAl–O	interface	0.90	−0.22 *	−0.08	0.00	0.02
	middle layer	0.88 (1.64)	−0.44	−0.16	0.00	0.00
TiCo–O	interface	0.76	−0.12	0.08 *	0.04	0.02
	middle layer	0.88 (1.62)	−0.08	0.12	0.00	0.00
CoCo–O	interface	0.18 *	0.32(0.14)	−0.06 *	0.00	0.00
	middle layer	0.80 (1.66)	−0.38	−0.12	0.00	0.00
TiTi–O	interface	0.32	0.10 *	−	0.00	0.00
	middle layer	0.88 (1.76)	−0.44	−0.12	0.00	0.00
AlAl–O	interface	1.46 *	0.56 *	−0.02	0.00	0.00
	middle layer	0.86 (1.70)	−0.40	−0.12	0.00	0.00

3.3. Interface Electronic Properties

In order to analyze the electronic properties of interface layer atoms, the densities of state (DOS) of the two outermost layer atoms in Heusler alloy Ti$_2$CoAl and the first interface layer atom in MgO for 10 potential terminations were analyzed in the Ti$_2$CoAl/MgO surpercell. Figure 2 shows that the atom-resolved DOS at the middle layer was extremely close to the feature of the bulk. In all 10 terminations, we could find that the spin-down gap in the bulk had been destroyed. In TiAl–Mg and AlAl–O termination, the spin-down gap narrowed down compared with the middle layers. In the rest of the eight terminations, some peaks from interface/subinterface Co or Ti atoms appeared in the spin-down gap and crossed the Fermi level. For the interface/subinterface Al atom, a slight spin polarization was observed at the Fermi level. In all 10 terminations, all interface Mg and O atoms suffered an extremely small spin polarization at the Fermi level, thereby indicating that the interface alloy–Mg and alloy–O bonding was not strong enough to contribute to the electronic properties at the Fermi level. Unfortunately, unexpected interface states appeared at the Fermi level and destroyed the "ideal" half-metallicity observed in the bulk. Meanwhile, the interface states in the AlAl–Mg, TiTi–Mg, TiCo–Mg and TiTi–O terminations had entirely filled the spin-metallic gaps in the spin-down channel at the Fermi level. By contrast, in the CoCo–Mg and CoCo–O terminations, the electronic structure was very similar to the behavior of the middle layer atoms. Evidence of high spin polarization has been previously reported in the CoCo termination of the Ti$_2$CoAl(100) surface system [34]. In the TiAl–Mg and AlAl–O terminations, the half-metallic gap suffered minimal destruction, and the DOSs were in accordance with that of the middle layer atoms. We deduced that a high spin polarization might be reserved in these terminations.

Given the significance of interface polarization in MTJs, we examined the interface spin polarizations in various terminations, especially for the several interface layers in contact with the MgO slab. Table 3 summarizes the spin polarization P, spin-up state density $N_\uparrow$, and spin-down state density $N_\downarrow$ at the Fermi level. The contributions from the interface and subinterface layers in Ti_2CoAl (labeled I-type) and from the three interface layers with the addition of the first interface layer in MgO (labeled II-type) were also analyzed. Surprisingly, the lowest value was less than 1% in three interface layers for the TiTi–Mg terminations for the TiCo–Mg, TiAl–Mg, and AlAl–Mg terminations. However, for the first two interface layers in the Ti_2CoAl slab, the P value was the largest. According to the Julliere formula, we could deduce that the TMR value was not larger than 150% in the $Ti_2CoAl/MgO(100)$ heterojunction at low temperature.

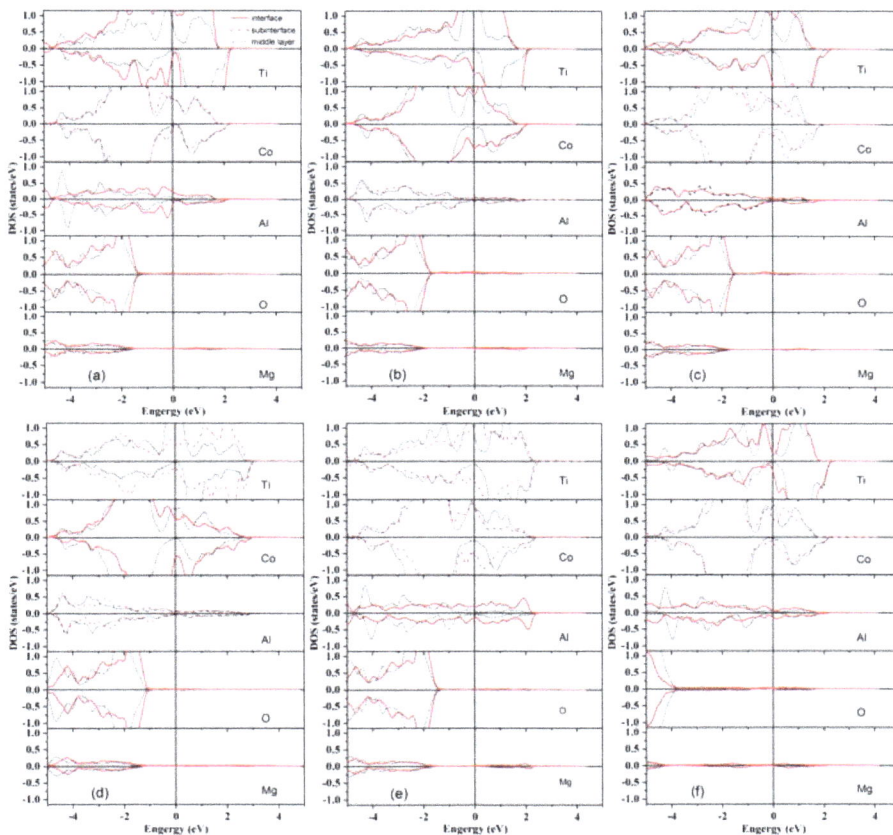

Figure 2. *Cont.*

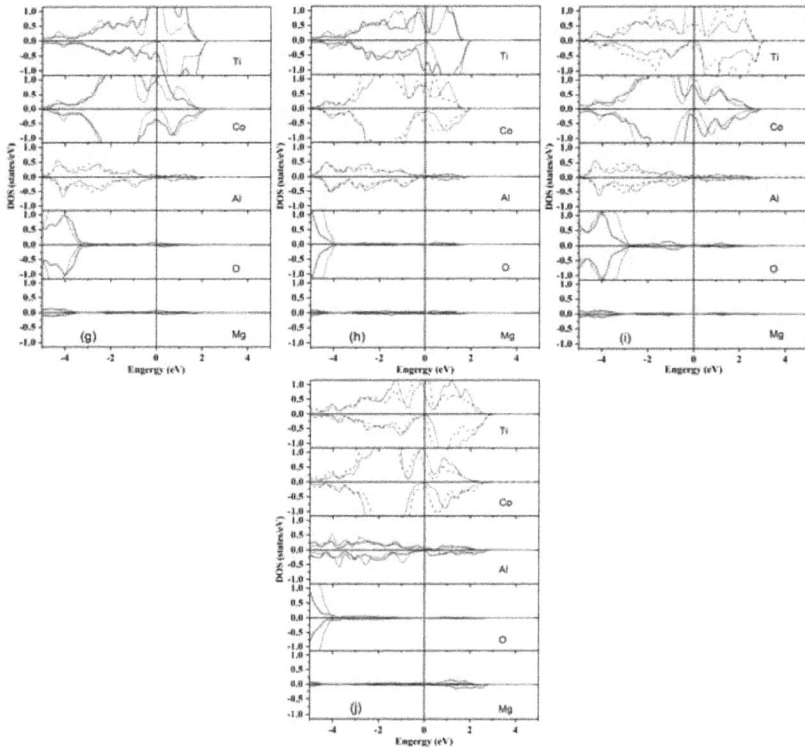

Figure 2. The densities of state (DOS) of the two outermost layer atoms in Heusler alloy Ti_2CoAl and the first interface layer atom in MgO for 10 potential terminations in Ti_2CoAl/MgO junctions. (a) TiAl–Mg termination; (b) TiCo–Mg termination; (c) TiTi–Mg termination; (d) CoCo–Mg termination; (e) AlAl–Mg termination; (f) TiAl–O termination; (g) TiCo–O termination; (h) TiTi–O termination; (i) CoCo–O termination; and (j) AlAl–O termination.

Table 3. Spin polarization P, spin-up state density $N_\uparrow$ at the Fermi level, and spin-down state density $N_\downarrow$ at the Fermi level. I-type includes the interface and subinterface layers in Ti_2CoAl, while the II-type comprises I-type layers and the interface layer in MgO.

Interface Layers	TiAl–Mg	TiCo–Mg	CoCo–Mg	TiTi–Mg	AlAl–Mg
I-type P (%)	66.21	20.16	8.22	−1.23	22.36
I-type $N_\uparrow$ (states/eV)	5.51	6.55	3.44	6.30	4.60
I-type $N_\downarrow$ (states/eV)	1.12	4.35	2.92	6.46	2.92
II-type P (%)	65.72	19.85	8.13	−0.77	22.25
II-type $N_\uparrow$ (states/eV)	5.59	6.61	3.46	6.42	4.67
II-type $N_\downarrow$ (states/eV)	1.16	4.42	2.94	6.52	2.97
Interface Layers	**TiAl–Mg**	**TiCo–Mg**	**CoCo–Mg**	**TiTi–Mg**	**AlAl–Mg**
I-type P (%)	30.51	54.62	46.50	−27.96	68.83
I-type $N_\uparrow$ (states/eV)	3.58	7.02	4.42	3.38	3.48
I-type $N_\downarrow$ (states/eV)	1.91	2.06	1.61	6.00	0.64
II-type P (%)	30.29	54.20	46.34	−27.27	65.20
II-type $N_\uparrow$ (states/eV)	3.72	7.51	4.58	3.47	3.61
II-type $N_\downarrow$ (states/eV)	1.99	2.23	1.68	6.07	0.76

4. Conclusions

Using the first-principle calculations within DFT, the interface structures, atom-resolved magnetism, atom-resolved DOS, and spin polarization of 10 atomic terminations in the $Ti_2CoAl/MgO(100)$ heterojunction were systematically examined. The results revealed that in equilibrium interface structures, the length of the alloy–Mg bond was much longer than that of the alloy–O bond because of the forceful repulsion interactions between the Heusler interface atoms and Mg atoms, especially those between metal atoms and Mg atoms. Owing to the competition among d-electron hybridization, d-electron localization originating from interface atomic bonding, and the moving effect of interface metal atoms, the interface atomic magnetic moments were complex and varied. In general, the interface atomic magnetism was slightly larger than the corresponding bulk-like layer atoms in alloy–Mg terminations. However, the opposite magnetic behavior was observed in alloy–O terminations. Analyzing the electronic properties near the Fermi level, we found that unexpected interface states appeared at the Fermi level for all terminations and destroyed the "ideal" half-metallicity observed in the bulk. A minimal destruction in the half-metallic gap and a maximal spin polarization of approximately 65% could only be observed in TiAl–Mg and AlAl–O terminations. From the Julliere formula, we could deduce that the TMR value was not larger than 150% in the $Ti_2CoAl/MgO(100)$ heterojunction at low temperature.

Author Contributions: Methodology, B.W.; software, G.Z. and X.W.; data curation, Y.F. and Y.C.; writing—original draft preparation, H.H.; writing—review and editing, B.W.

Funding: This work was partly supported by the National Natural Science Foundation of China (11304410), Key Laboratory and Scientific Research Foundation of Zunyi City (SSKH [2015] 55), Natural Science Foundation of Technology Department (QKHJZ-LKZS [2014] 10, QJHJZ-LKZS [2012]03, and KHJZ[2014]2170), Youth Science Foundation of Education Ministry (QJHKZ [2012] 084, QJHKY[2018]310), and the Key Support Discipline ([2011]275) of Guizhou Province of China.

Conflicts of Interest: The authors declare no conflict of interest.

References

1. Segu, D.Z.; Khan, P.V.; Hwang, P. Experimental and direct numerical analysis of hard-disk drive. *J. Mech. Sci. Technol.* **2018**, *32*, 3507–3513. [CrossRef]
2. Kubota, T.; Ina, Y.; Wen, Z.; Takanashi, K. Interface Tailoring Effect for Heusler Based CPP-GMR with an $L1_2$-Type Ag_3Mg Spacer. *Materials* **2018**, *11*, 219. [CrossRef] [PubMed]
3. Satoshi, S.; Susumu, H.; Masayuki, T.; Yuzo, K.; Hitoshi, I. All-metallic nonlocal spin valves using polycrystalline $Co_2(FeMn)Si$ Heusler alloy with large output. *Appl. Phys. Express* **2015**, *8*, 023103.
4. Koki, M.; Shinya, K.; Yukiko, K.T.; Kouta, K.; Yoshichika, O.; Seiji, M.; Kazuhiro, H. High output voltage of magnetic tunnel junctions with a $Cu(In_{0.8}Ga_{0.2})Se_2$ semiconducting barrier with a low resistance–area product. *Appl. Phys. Express* **2017**, *10*, 013008.
5. Li, S.; Takahashi, Y.K.; Sakuraba, Y.; Chen, J.; Furubayashi, T.; Mryasov, O.; Faleev, S.; Hono, K. Current-perpendicular-to-plane giant magnetoresistive properties in $Co_2Mn(Ge_{0.75}Ga_{0.25})/Cu_2TiAl/Co_2Mn(Ge_{0.75}Ga_{0.25})$ all-Heusler alloy pseudo spin valve. *J. Appl. Phys.* **2016**, *119*, 093911. [CrossRef]
6. Çakır, A.; Acet, M. Non-volatile high-temperature shell-magnetic pinning of Ni-Mn-Sn Heusler precipitates obtained by decomposition under magnetic field. *J. Magn. Magn. Mater.* **2018**, *448*, 13–18. [CrossRef]
7. Nayak, A.K.; Kumar, V.; Ma, T.; Werner, P.; Pippel, E.; Sahoo, R.; Damay, F.; Rößler, U.K.; Felser, C.; Parkin, S.S.P. Magnetic antiskyrmions above room temperature in tetragonal Heusler materials. *Nature* **2017**, *548*, 561–566. [CrossRef] [PubMed]
8. Singh, S.; D'souza, S.W.; Nayak, J.; Suard, E.; Chapon, L.; Senyshyn, A.; Petricek, V.; Skourski, Y.; Nicklas, M.; Felser, C.; et al. Room-temperature tetragonal non-collinear Heusler antiferromagnet Pt_2MnGa. *Nat. Commun.* **2016**, *7*, 12671. [CrossRef] [PubMed]
9. Cai, Z.; Fenglong, W.; Gesang, D.; Jinli, Y.; Changjun, J. Piezostrain tuning non-volatile 90° magnetic easy axis rotation in Co_2FeAl Heusler alloy film grown on $Pb(Mg_{1/3}Nb_{2/3})O_3$-$PbTiO_3$ heterostructures. *J. Phys. D Appl. Phys.* **2016**, *49*, 455001.

10. Dutta, S.; Nikonov, D.E.; Manipatruni, S.; Young, I.A.; Naeemi, A. Overcoming thermal noise in non-volatile spin wave logic. *Sci. Rep.* **2017**, *7*, 1915. [CrossRef] [PubMed]

11. Julliere, M. Tunneling between ferromagnetic films. *Phys. Lett. A* **1975**, *54*, 225–226. [CrossRef]

12. Chen, Y.; Wu, B.; Yuan, H.; Feng, Y.; Chen, H. The defect-induced changes of the electronic and magnetic properties in the inverse Heusler alloy Ti$_2$CoAl. *J. Solid State Chem.* **2015**, *221*, 311–317. [CrossRef]

13. Vasileiadis, T.; Waldecker, L.; Foster, D.; Da Silva, A.; Zahn, D.; Bertoni, R.; Palmer, R.E.; Ernstorfer, R. Ultrafast heat flow in heterostructures of Au nanoclusters on thin-films: Atomic-disorder induced by hot electrons. *arXiv*, 2018; arXiv:1803.00074. [CrossRef] [PubMed]

14. Bo, W.; Hongkuan, Y.; Anlong, K.; Yu, F.; Hong, C. Tunable magnetism and half-metallicity in bulk and (100) surface of quaternary Co$_2$MnGe$_{1-x}$Ga$_x$ Heusler alloy. *J. Phys. D Appl. Phys.* **2011**, *44*, 405301.

15. Shen, X.; Yu, G.; Zhang, C.; Wang, T.; Huang, X.; Chen, W. A theoretical study on the structures and electronic and magnetic properties of new boron nitride composite nanosystems by depositing superhalogen Al$_{13}$ on the surface of nanosheets/nanoribbons. *Phys. Chem. Chem. Phys.* **2018**, *20*, 15424–15433. [CrossRef] [PubMed]

16. Yang, G.; Li, D.; Wang, S.; Ma, Q.; Liang, S.; Wei, H.; Han, X.; Hesjedal, T.; Ward, R.; Kohn, A. Effect of interfacial structures on spin dependent tunneling in epitaxial L10-FePt/MgO/FePt perpendicular magnetic tunnel junctions. *J. Appl. Phys.* **2015**, *117*, 083904. [CrossRef]

17. Xu, A.; Shi, L.; Zhao, T. Thermal effects on the sedimentation behavior of elliptical particles. *Int. J. Heat Mass Tran.* **2018**, *126*, 753–764. [CrossRef]

18. Grimm, R.; Marchi, S. Direct thermal effects of the Hadean bombardment did not limit early subsurface habitability. *Earth Planet. Sci. Lett.* **2018**, *485*, 1–8. [CrossRef]

19. Parkin, S.S.; Kaiser, C.; Panchula, A.; Rice, P.M.; Hughes, B.; Samant, M.; Yang, S.-H. Giant tunnelling magnetoresistance at room temperature with MgO (100) tunnel barriers. *Nat. Mater.* **2004**, *3*, 862–867. [CrossRef] [PubMed]

20. Ozawa, E.; Tsunegi, S.; Oogane, M.; Naganuma, H.; Ando, Y. The effect of inserting thin Co$_2$MnAl layer into the Co$_2$MnSi/MgO interface on tunnel magnetoresistance effect. *J. Phys. Conf. Ser.* **2011**, *266*, 012104. [CrossRef]

21. Liu, H.-X.; Honda, Y.; Taira, T.; Matsuda, K.-I.; Arita, M.; Uemura, T.; Yamamoto, M. Giant tunneling magnetoresistance in epitaxial Co$_2$MnSi/MgO/Co$_2$MnSi magnetic tunnel junctions by half-metallicity of Co$_2$MnSi and coherent tunneling. *Appl. Phys. Lett.* **2012**, *101*, 132418. [CrossRef]

22. Bai, Z.; Lu, Y.; Shen, L.; Ko, V.; Han, G.; Feng, Y. Transport properties of high-performance all-Heusler Co$_2$CrSi/Cu$_2$CrAl/Co$_2$CrSi giant magnetoresistance device. *J. Appl. Phys.* **2012**, *111*, 093911. [CrossRef]

23. Rotjanapittayakul, W.; Prasongkit, J.; Rungger, I.; Sanvito, S.; Pijitrojana, W.; Archer, T. Search for alternative magnetic tunnel junctions based on all-Heusler stacks. *arXiv*, 2018; arXiv:1805.08603. [CrossRef]

24. Graf, T.; Felser, C.; Parkin, S.S. Simple rules for the understanding of Heusler compounds. *Prog. Solid State Chem.* **2011**, *39*, 1–50. [CrossRef]

25. Taira, T.; Ishikawa, T.; Itabashi, N.; Matsuda, K.-I.; Uemura, T.; Yamamoto, M. Spin-dependent tunnelling characteristics of fully epitaxial magnetic tunnel junctions with a Heusler alloy Co$_2$MnGe thin film and a MgO barrier. *J. Phys. D Appl. Phys.* **2009**, *42*, 084015. [CrossRef]

26. Furubayashi, T.; Kodama, K.; Sukegawa, H.; Takahashi, Y.; Inomata, K.; Hono, K. Current-perpendicular-to-plane giant magnetoresistance in spin-valve structures using epitaxial Co$_2$FeAl$_{0.5}$Si$_{0.5}$/Ag/Co$_2$FeAl$_{0.5}$Si$_{0.5}$ trilayers. *Appl. Phys. Lett.* **2008**, *93*, 122507. [CrossRef]

27. Katsnelson, M.; Irkhin, V.Y.; Chioncel, L.; Lichtenstein, A.; De Groot, R.A. Half-metallic ferromagnets: From band structure to many-body effects. *Rev. Mod. Phys.* **2008**, *80*, 315. [CrossRef]

28. Zhang, X.Y.; Guo, Q.; Li, Y.D.; Wen, L. Total ionizing dose and synergistic effects of magnetoresistive random-access memory. *Nucl. Sci. Tech.* **2018**, *29*, 111. [CrossRef]

29. Bayar, E.; Kervan, N.; Kervan, S. Half-metallic ferrimagnetism in the Ti$_2$CoAl Heusler compound. *J. Magn. Magn. Mater.* **2011**, *323*, 2945–2948. [CrossRef]

30. Drief, M.; Guermit, Y.; Benkhettou, N.; Rached, D.; Rached, H.; Lantri, T. First-Principle Study of Half-Metallic Ferrimagnet Behavior in Titanium-Based Heusler Alloys Ti$_2$FeZ (Z = Al, Ga, and In). *J. Supercond. Nov. Magn.* **2018**, *31*, 1059–1065. [CrossRef]

31. Dahmane, F.; Benalia, S.; Djoudi, L.; Tadjer, A.; Khenata, R.; Doumi, B.; Aourag, H. First-principles study of structural, electronic, magnetic and half-metallic properties of the Heusler alloys Ti_2ZAl (Z = Co, Fe, Mn). *J. Supercond. Nov. Magn.* **2015**, *28*, 3099–3104. [CrossRef]

32. Sterwerf, C.; Meinert, M.; Schmalhorst, J.-M.; Reiss, G. High TMR ratio in Co_2FeSi and Fe_2CoSi based magnetic tunnel junctions. *arXiv*, 2013; arXiv:1308.2072.

33. Chen, Y.; Wu, B.; Feng, Y.; Yuan, H.-K.; Chen, H. Half-metallicity and magnetism of the quaternary inverse full-Heusler alloy $Ti_{2-x}M_xCoAl$ (M = Nb, V) from the first-principles calculations. *Eur. Phys. J. B* **2014**, *87*, 24. [CrossRef]

34. Feng, Y.; Wu, B.; Yuan, H.; Kuang, A.; Chen, H. Magnetism and half-metallicity in bulk and (100) surface of Heusler alloy Ti_2CoAl with Hg_2CuTi-type structure. *J. Alloys Compd.* **2013**, *557*, 202–208. [CrossRef]

35. Perdew, J.P.; Burke, K.; Ernzerhof, M. Generalized gradient approximation made simple. *Phys. Rev. Lett.* **1996**, *77*, 3865–3868. [CrossRef] [PubMed]

36. Vanderbilt, D. Soft self-consistent pseudopotentials in a generalized eigenvalue formalism. *Phys. Rev. B* **1990**, *41*, 7892. [CrossRef]

37. Filippov, S.; Magadov, K.Y. Spin polarization-scaling quantum maps and channels. *Lobachevskii J. Math.* **2018**, *39*, 65–70. [CrossRef]

38. Hu, Y.; Zhang, J.-M. First-principles study on the thermodynamic stability, magnetism, and half-metallicity of full-Heusler alloy Ti_2FeGe (001) surface. *Phys. Lett. A* **2017**, *381*, 1592–1597. [CrossRef]

39. Wang, Y.X.; Xia, T.S. Spin Injection into Graphene from Heusler Alloy Co_2MnGe (111) Surface: A First Principles Study. *Mater. Sci. Forum* **2018**, *914*, 111–116. [CrossRef]

applied
sciences

MDPI

Article

Spin Gapless Semiconductor–Nonmagnetic Semiconductor Transitions in Fe-Doped Ti$_2$CoSi: First-Principle Calculations

Yu Feng [1], Zhou Cui [1], Ming-sheng Wei [1], Bo Wu [2,3,*] and Sikander Azam [4]

1   School of Physics and Electronic Engineering, Jiangsu Normal University, Xuzhou 221116, China; fengyu@jsnu.edu.cn (Y.F.); zhoucuijsnu@163.com (Z.C.); weims@jsnu.edu.cn (M.-s.W.)
2   School of Physics and Electronic Science, Zunyi Normal University, Zunyi 563002, China
3   School of Marine Science and Technology, Northwestern Polytechnical University, Xi'an 710072, China
4   Department of Physics, The University of Lahore, Sargodha Campus, Sargodha 40100, Pakistan; sikander.physicst@gmail.com
*   Correspondence: fqwubo@zync.edu.cn; Tel.: +86-0851-2892-7153

Received: 28 October 2018; Accepted: 6 November 2018; Published: 9 November 2018

Abstract: Employing first-principle calculations, we investigated the influence of the impurity, Fe atom, on magnetism and electronic structures of Heusler compound Ti$_2$CoSi, which is a spin gapless semiconductor (SGS). When the impurity, Fe atom, intervened, Ti$_2$CoSi lost its SGS property. As Ti^A atoms (which locate at (0, 0, 0) site) are completely occupied by Fe, the compound converts to half-metallic ferromagnet (HMF) TiFeCoSi. During this SGS→HMF transition, the total magnetic moment linearly decreases as Fe concentration increases, following the Slate–Pauling rule well. When all Co atoms are substituted by Fe, the compound converts to nonmagnetic semiconductor Fe$_2$TiSi. During this HMF→nonmagnetic semiconductor transition, when Fe concentration y ranges from y = 0.125 to y = 0.625, the magnetic moment of Fe atom is positive and linearly decreases, while those of impurity Fe and Ti^B (which locate at (0.25, 0.25, 0.25) site) are negative and linearly increase. When the impurity Fe concentration reaches up to y = 1, the magnetic moments of Ti, Fe, and Si return to zero, and the compound is a nonmagnetic semiconductor.

Keywords: Heusler alloy; electronic structure; magnetism; doping

---

1. Introduction

As one of the most outstanding material classes, Heusler compounds with a chemical formula of X$_2$YZ are a large family containing more than 1500 members [1,2]. When the number of valence electrons (VEs) of X atom is more than that of Y atom, Heusler compounds are known to be of a conventional type, i.e., Cu$_2$MnAl type with a space group of FM-3M, see Figure 1a [3]. While, when the number of VEs of X atom is less than that of Y atom, Heusler compounds crystallize in an inverse type, i.e., Hg$_2$CuTi type with space group of F-43M, see Figure 1b [4]. In addition, when two X atoms are different, the chemical formula of Heusler compounds converts to XX'YZ, and it is a quaternary type, i.e., LiMgPbSn type with space group of F-43M, see Figure 1c [5,6]. Owing to the fact that there is huge number of Heusler compounds that could be comprised by a combination of different elements, Heusler compounds exhibit diverse properties. Several Heusler compounds, such as Co$_2$MnSi, Ti$_2$CoAl, CoFeMnAl, and others, have been theoretically predicted and experimentally confirmed to be half-metallic ferromagnets [7–13]. Due to the special band structure, that the majority of the band shows metallicity, while an energy gap exists in the minority band, half-metallic Heusler compounds could offer theoretically 100% spin-polarized current. Besides, most of the half-metallic Heusler compounds possess high Curie temperature, and their lattice constants are very close to many semiconductors, such as MgO and GaAs. They are, therefore, regarded as one of the most excellent candidates for electrode materials of spintronics devices, such as magnetic tunnel junctions (MTJ) and current-perpendicular-to-plane spin valves (CPP-SV), and have a great application potential in magnetic random access memory (MRAM), ultra-high-speed reading in magnetic read heads of hard disk drivers (HDD) and spin transfer torque (STT) devices in spin random access memory [14–21]. In addition, a lot of Heusler compounds, especially half-Heusler compounds with chemical formula of XYZ, exhibit semiconductor character. Owing to high Seebeck coefficient, large electrical conductivity, good thermal stability, and environmentally friendly constituents, semiconducting Heusler compounds, such as *n*-type MNiSn (M = Ti, Zr, Hf) and *p*-type ErNiSn, HfPtSn, became promising thermoelectric materials which could recycle waste heat into electricity [22–25]. Hence, they are great useful for wearable devices like smart watches, and for sensors in industrial process monitoring. More recently, much attention has been paid to a new subfamily of Heusler compounds which were characterized with a novel band structure, that there is an energy gap that lies in minority bands, while the valence and conduction band edges of the majority of electrons touch at Fermi level, resulting in a zero-width gap. They are, therefore, classified to be spin gapless semiconductors (SGSs) [26]. Heusler compound Mn_2CoAl with SGS properties has been successfully fabricated, and high Curie temperature (T_C) of 720 K as well as magnetism of 2 μ_B were detected [27–29]. Others, like Ti_2CoSi and CoFeCrGa, also received intense research interest [30–34]. Owing to the extraordinary band structure, both electrons and holes of SGS Heusler compounds can be spin-polarized, and almost no threshold energy is required to move electrons from the valence band to the conduction band, as the mobility of carriers is stronger than that in regular semiconductors. Hence, they are considered to be possible candidates to substitute for diluted magnetic semiconductors (DMS) [35]. Due to the reason that diverse valence electrons configurations of Heusler compounds result in varied magnetic properties and electronic structures, in this paper, we studied the influence of impurity Fe atom on magnetism and electronic structures of Heusler compound Ti_2CoSi.

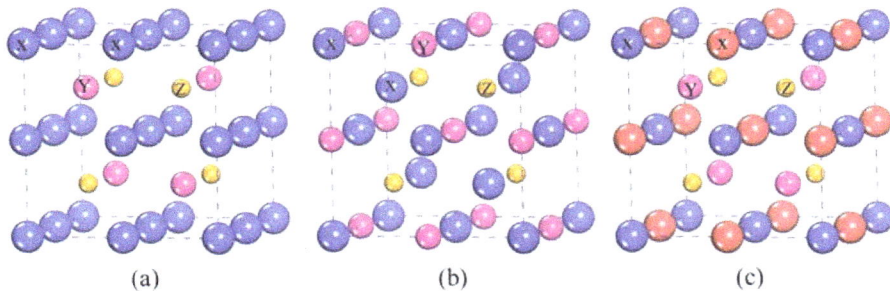

Figure 1. Schematic representation of (**a**) Cu_2MnAl-type Heusler compound; (**b**) Hg_2CuTi-type Heusler compound; (**c**) LiMnPbSn-type Heusler compound.

2. Structures and Calculation Methods

All calculations are performed by employing the VASP Package based on the density functional theory (DFT) [36]. Plane-wave basis sets, together with the projector-augmental wave (PAW) [37] method, are chosen to deal with electron–ion interaction. The valence-electron configurations of Ti ($3d^24s^2$), Fe ($3d^64s^2$), Co ($3d^74s^2$), and Si ($3s^23p^2$) are selected. The $7 \times 7 \times 7$ mesh of special k-points in the Brillouin zone is applied. In the self-consistent calculation, we select the refined 1×10^{-6} eV/atom as the SCF convergence criterion, and 360 eV as energy cutoff, respectively. When the positions of atoms are relaxed, we set a convergence criterion of 0.02 eV/Å. All structures are built with $2 \times 1 \times 1$ supercell. For the doped compound calculations, we geometrically optimize all supercells by using the same parameters employed in the bulk calculation. All technical parameters have been tested carefully to ensure the accuracy of the results.

3. Results and Discussion

Interface Structures

The electronic structure of Ti_2CoSi is firstly calculated. In Figure 2, there is a large energy gap of 0.671 eV exists in spin down band, and the Fermi level locates at the top of the gap. Such an energy gap is a result of exchange splitting between spin down unoccupied antibonding bands (which are localized at Co, Ti^A and Ti^B atoms) and spin down occupied bands (which are predominantly of Co character). As for spin up band, it exhibits an obvious zero-width gap around the Fermi level. It also can be seen from Figure 3 that the maximum of the valence band sits at Γ point, while the minimum of the conduction bands locates at X point; such a closed spin up gap, therefore, is an indirect gap.

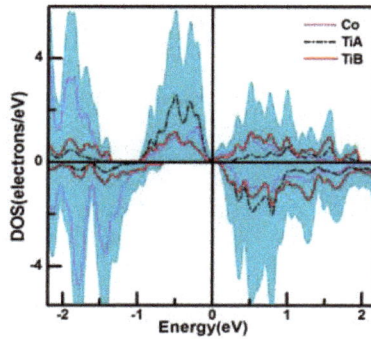

Figure 2. The electronic structure of Ti_2CoSi.

When Fe atom is introduced into Ti_2CoSi, it could lead to four doping structures, thus, X(Fe) doping structure where X atom (X = Ti^A, Ti^B, Co, Si) is substituted by an Fe atom. In order to determine which doping structure is more favorable, the formation energy is calculated by the following equation $E_f = E' - E - \sum_i n_i \mu_i$, where E_f is formation energy, and E' and E are total energy of the doping structure and undoping structure, respectively. The integer n_i is the number of atoms that has been removed from (n_i is negative value) or added to (n_i is positive value) to form the disorders, and μ_i is the corresponding chemical potential which represents the energy of the reservoirs. According to our calculation, the highest E_f of 1.028 eV belongs to Ti^B(Fe) doping structure, and that of Co(Fe) and Si(Fe) are also as high as -0.023 eV and -0.018 eV, respectively, while the minimal E_f of -0.927 eV occurs in Ti^A(Fe) doping structure. It reveals that Ti^B, Co, and Si atoms are hard to be replace by Fe atom, however, Ti^A atom could be easily replaced with Fe atom. Therefore, we focus on the $Ti^A_{1-x}Fe_xTi^BCoSi$ doping structure, where the doping concentration x = 0, 0.125, 0.25, 0.375, 0.5, 0.625, 0.75, 0.875, and 1.

Appl. Sci. **2018**, *8*, 2200

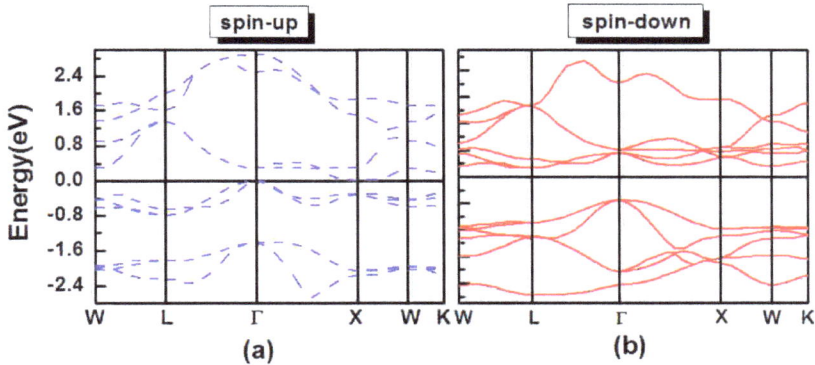

Figure 3. The band structure of Ti$_2$CoSi. (**a**) Spin-up band structure. (**b**) Spin-down band structure.

In Figure 4, when the Fe atom intervenes in Ti$_2$CoSi and the doping concentration increases, the zero-width gap in spin up states is destroyed, and the spin up band strides over the Fermi level and shows metallic behavior. On the other hand, doping Fe atom produces a negative impact on the spin down energy gap, causing the width of the gap to vanish, and the Fermi level drops into a spin down conduction band. As a result, the SGS character of Ti$_2$CoSi is completely destroyed by the impurity Fe atom. When the doping concentration x = 0.875, the top of the spin down valence band and the bottom of spin down conduction band touch at Fermi level, and form an indirect closed spin down gap. The doping structure, Ti$^A_{0.125}$Fe$_{0.875}$Ti^{B}CoSi, is a gapless half-metal. As the doping concentration increases up to x = 1, the closed spin down gap is opened, while the spin up states still cross the Fermi level. Therefore, when x = 1, the structure converts to a quaternary Heusler alloy TiFeCoSi with half-metallic character.

It can be seen from Figure 5 that TiFeCoSi possesses a wide half-metallic energy gap of about 0.56 eV, and the Fermi level is located slightly above the middle of the gap and, hence, it is predicted to have stable half-metallicity. Both spin down occupied bonding bands and spin down unoccupied antibonding bands are mainly localized by Co and Fe atoms. Figure 6 shows the half-metallicity of TiFeCoSi as a function of the lattice constant, and it holds the half-metallic energy gap when its lattice constant increases from 5.4 Å to 6.1 Å. When the lattice constant is 5.4 Å, the Fermi level lies at the top of spin down valence band, and when TiFeCoSi is further compressed, the Fermi level would drop into the valence band and lose its half-metallicity. With the lattice constant increases, the Fermi level gradually moves from a low energy zone to a high energy zone. As the lattice constant increases up to 6.1 Å, the Fermi level locates at the bottom of spin down conduction band, and when TiFeCoSi is further stretched, the Fermi level would move into the conduction band, and half-metallicity is also destroyed.

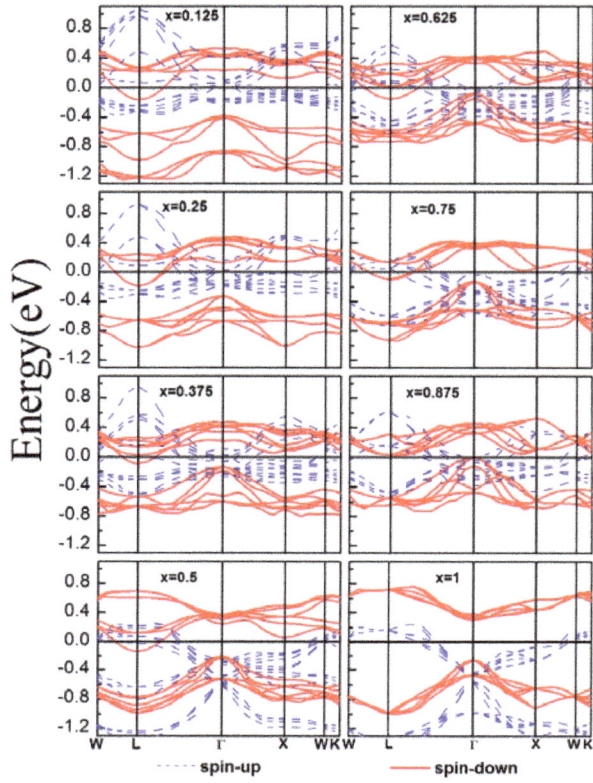

Figure 4. The band structures of $Ti^A_{1-x}Fe_xTi^BCoSi$.

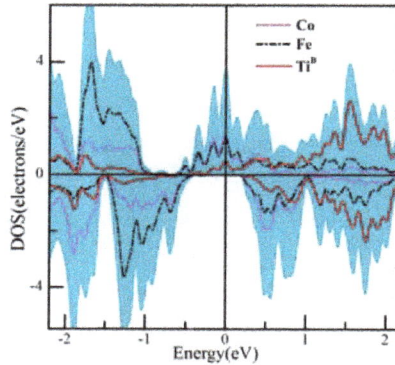

Figure 5. The electronic structure of TiFeCoSi.

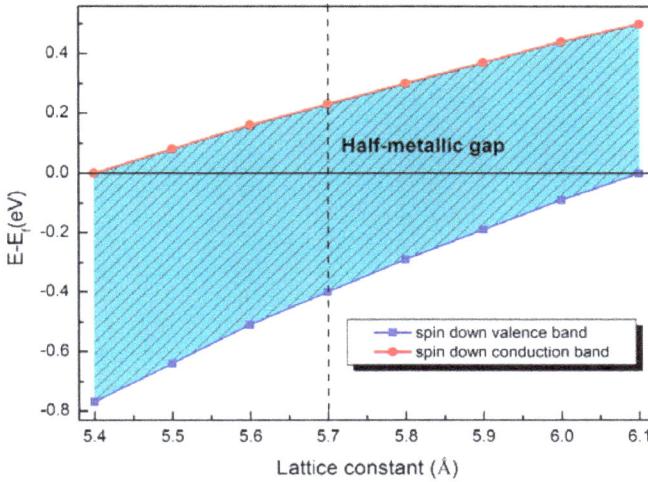

Figure 6. Dependence of the half-metallic state on the lattice constant. The dotted line indicates the half-metallic state at equilibrium lattice constant.

The atom-resolved spin magnetic moment (AMM) and total magnetic moment (TMM) of Fe doped Ti_2CoSi as a function of Fe concentration is shown in Figure 7. As for pure Ti_2CoSi, Ti^A and Ti^B contribute the main part of TMM, while Co offers a small part, due to the reason that low valence transition metal Ti atom has larger spin splitting than the high valence transition metal Co. When Fe atom doping in Ti_2CoSi forms $Ti^A_{1-x}Fe_xTi^BCoSi$, the contribution of Fe atom to the TMM is lower than Ti^A atom. As the Fe doping concentration increases, the AMM of Ti^A, Ti^B, and Fe atoms lineally decline, while AMM of the Co atom is enhanced, and it is almost as high as Fe atom when the doping concentration increases to x = 1. It also can be seen that when x ranges from 0 to 0.625, the AMMs of Ti^A, Ti^B, Fe, and Co are positive values, indicating the ferromagnetic arrangement among these transition metal elements. While, when x > 0.625, the AMM of Ti^B atom reverses to a negative value, leading to the fact that the doped structure changes into a ferrimagnet. Overall, the TMM lineally decreases with Fe concentration increases, and it is in good agreement with the Slater–Pauling rule, which can be written as TMM = Z − 24, where Z is the total number of valence electrons.

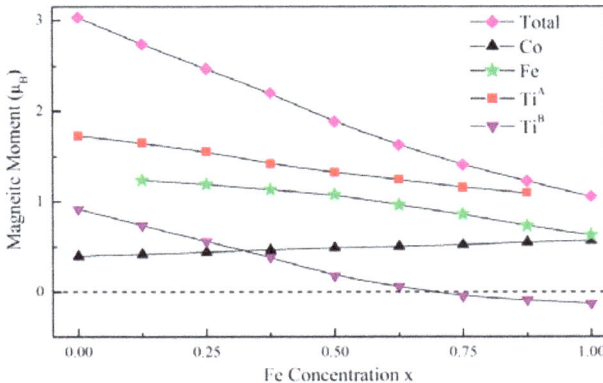

Figure 7. The calculated total magnetic moment and atom-resolved spin magnetic moments (AMMs) of the Co, Fe, Ti^A, and Ti^B atoms of $Ti^A_{1-x}Fe_xTi^BCoSi$ as a function of impurity Fe concentration x.

Based on the half-metallic TiFeCoSi, we continue to introduce the impurity, Fe atom, and impurity Fe may occupy Ti, Co, or Si atoms, forming a Ti(Fe), Co(Fe), or Si(Fe) doping structure, respectively. The formation energy of these three possible doping structures are also calculated by Equation (1). Si(Fe) and Ti(Fe) doping structures respectively exhibit high formation energies of 1.895 eV and 1.437 eV, revealing that Si and Ti are hard to be replaced by Fe atom. While Co(Fe) doping structure possesses the minimal formation energy of −0.153 eV, indicating that the Co atom is easily occupied by Fe atom and forms the $TiFeCo_{1-y}Fe_ySi$ doping structure, where y is the concentration of the impurity Fe atom, and y = 0.125, 0.25, . . . , 0.875, 1. Figure 8 shows the band structures of $TiFeCo_{1-y}Fe_ySi$. When impurity Fe concentration y ranges from 0.125 to 0.625, spin up bands stride over the Fermi level, while there is an obvious energy gap that exists in the spin down bands. Therefore, $TiFeCo_{1-y}Fe_ySi$ maintains its half-metallicity when the impurity Fe concentration increases from 0.125 to 0.625. As y increases to 0.75, the Fermi level drops into the spin down conduction band, and the doping structure loses its half-metallicity, and the spin polarization decreases to about 74%. In addition, with an increase of the impurity Fe concentration, the overlap degree between spin up valence band and spin down valence band increases. When y reaches up to 0.875, the Fermi level still exists in the spin down conduction band, and there is much overlap between the spin up and spin down bands, and the energies of the spin up and spin down bands are almost the same, making the spin polarization seriously drop to only 3%. Hence, the doping structure $TiFeCo_{0.125}Fe_{0.875}Si$ can be regarded as pseudo-semiconductor. As the impurity Fe concentration further increases up to y = 1, the spin up and spin down bands completely overlap, and there is an energy gap of 0.418 eV that exists in both spin up and spin down band, and the structure Fe_2TiSi therefore converts to a semiconductor.

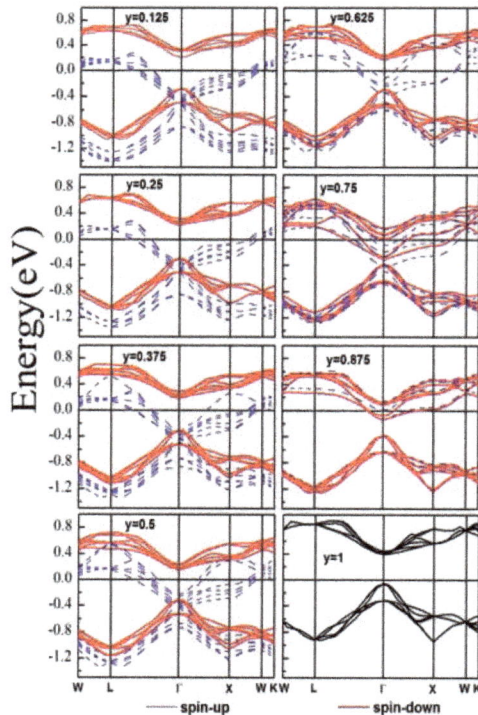

Figure 8. The band structures of $TiFeCo_{1-y}Fe_ySi$.

The AMM and TMM of Fe doped TiFeCoSi as a function of impurity Fe concentration is exhibited in Figure 9. When impurity Fe concentration increases from 0.125 to 0.75, AMM of Co atom has a slight increase, while that of the Fe atom descends obviously. In addition, the absolute value of AMMs of Ti^B and impurity Fe are weakened. It should be noted that when the impurity Fe concentration y increases to 0.875, the AMM of Co atom suffers a sharp decline, and AMMs of all atoms are extremely close to zero, owing to the fact that TiFeCo$_{0.125}$Fe$_{0.875}$Si shows a pseudo-semiconductor property. Furthermore, AMMs of Co and Fe are positive values, while that of Ti^B and impurity Fe are negative values, indicating that doping structure TiFeCo$_{1-y}$Fe$_y$Si is a ferrimagnet when y ranges from 0.125 to 0.75. As the impurity Fe concentration further increases up to y = 1, AMMs of all atoms are zero. As a result, Fe$_2$TiSi is a nonmagnetic semiconductor.

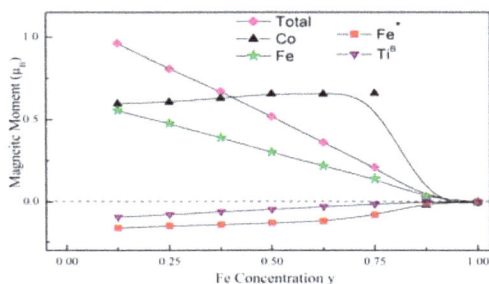

Figure 9. The calculated total magnetic moment and atom-resolved spin magnetic moments (AMMs) of the Co, Fe, Ti^B, and impurity Fe (Fe*) atoms of the TiFeCo$_{1-y}$Fe$_y$Si as a function of impurity Fe concentration y.

4. Conclusions

We investigated the influence of impurity Fe atom on magnetism and electronic structures of Heusler compound Ti$_2$CoSi by using the first-principle calculations within density functional theory (DFT). Ti$_2$CoSi is a spin gapless semiconductor (SGS) with an indirect closed spin down energy gap. It lost its SGS property when the impurity Fe atom intervened, and when the concentration of impurity Fe atom increases to x = 0.875, the compound shows gapless metal character. At the doping concentration of x = 1, Ti^A atoms are completely occupied by impurity Fe atoms, and the compound converts to TiFeCoSi, which is a half-metallic ferromagnet (HMF). During this SGS→HMF transition, the total magnetic moment linearly decreases with the concentration of impurity Fe atom increasing, which follows the Slate–Pauling rule well. When the impurity Fe further increases from y = 0.125 to y = 0.625, the doping compounds maintain their half-metallicity. While, when the doping concentration increases up to y = 1, the compound converts to semiconductor Fe$_2$TiSi. During this HMF→nonmagnetic semiconductor transition, when the concentration of impurity Fe atom ranges from y = 0.125 to y = 0.625, the magnetic moment of Fe atom is positive and linearly decreases, while that of impurity Fe and Ti^B are negative and linearly increase. When all Co atoms are substituted by Fe atoms, the magnetic moments of Ti, Fe, and Si return to zero, and the compound, therefore, is a nonmagnetic semiconductor.

Author Contributions: Methodology, Z.C.; software, M.-s.W. and S.A.; writing—original draft preparation, B.W.; writing—review and editing, Y.F.

Funding: This work was partly supported by the National Natural Science Foundation of China (11747114), Doctor Foundation of Jiangsu Normal University (16XLR022).

Conflicts of Interest: The authors declare no conflict of interest.

References

1. Nayak, A.K.; Nicklas, M.; Chadov, S.; Khuntia, P.; Shekhar, C.; Kalache, A.; Baenitz, M.; Skourski, Y.; Guduru, V.K.; Puri, A.; et al. Design of compensated ferrimagnetic Heusler alloys for giant tunable exchange bias. *Nat. Mater.* **2015**, *14*, 679–684. [CrossRef] [PubMed]

2. Graf, T.; Felser, C.; Parkin, S.S.P. Simple rules for the understanding of Heusler compounds. *Prog. Solid State Chem.* **2011**, *39*, 1–50. [CrossRef]

3. Galanakis, I.; Dederichs, P.H.; Papanikolaou, N. Slater-Pauling behavior and origin of the half-metallicity of the full-Heusler alloys. *Phys. Rev. B* **2002**, *66*, 174429. [CrossRef]

4. Skaftouros, S.; Özdoğan, K.; Şaşioğlu, E.; Galanakis, I. Generalized Slater-Pauling rule for the inverse Heusler compounds. *Phys. Rev. B* **2013**, *87*, 024420. [CrossRef]

5. Feng, Y.; Chen, H.; Yuan, H.; Zhou, Y.; Chen, X. The effect of disorder on electronic and magnetic properties of quaternary Heusler alloy CoFeMnSi with LiMgPbSb-type structure. *J. Magn. Magn. Mater.* **2015**, *378*, 7–15. [CrossRef]

6. Özdoğan, K.; Şaşioğlu, E.; Galanakis, I. Slater-Pauling behavior in LiMgPdSn-type multifunctional quaternary Heusler materials: Half-metallicity, spin-gapless and magnetic semiconductors. *J. Appl. Phys.* **2013**, *113*, 193903. [CrossRef]

7. Jourdan, M.; Minár, J.; Braum, J.; Kronenerg, A.; Chadov, S.; Balke, B.; Gloskovskii, A.; Kolbe, M.; Elmers, H.J.; Schönhense, G.; et al. Direct observation of half-metallicity in the Heusler compound Co2MnSi. *Nat. Commun.* **2014**, *5*, 3974. [CrossRef] [PubMed]

8. Feng, Y.; Xu, X.; Cao, W.; Zhou, T. Investigation of cobalt and silicon co-doping in quaternary Heusler alloy NiFeMnSn. *Comput. Mater. Sci.* **2018**, *147*, 251–257. [CrossRef]

9. Wang, L.; Jin, Y. A spin-gapless semiconductor of inverse Heusler Ti2CrSi alloy: First-principles prediction. *J. Magn. Magn. Mater.* **2015**, *385*, 55–59. [CrossRef]

10. Han, J.; Gao, G. Large tunnel magnetoresistance and temperature-driven spin filtering effect based on the compensated ferrimagnetic spin gapless semiconductor Ti2MnAl. *Appl. Phys. Lett.* **2018**, *113*, 102402. [CrossRef]

11. Gao, G.Y.; Hu, L.; Yao, K.L.; Luo, B.; Liu, N. Large half-metallic gaps in the quaternary Heusler alloys CoFeCrZ (Z = Al, Si, Ga, Ge): A first-principles study. *J. Alloy. Compd.* **2013**, *551*, 539–543. [CrossRef]

12. Felser, C.; Wollmann, L.; Chadov, S.; Fecher, G.H.; Parkin, S.S.P. Basics and prospective of magnetic Heusler compounds. *APL Mater.* **2015**, *3*, 041518. [CrossRef]

13. Feng, Y.; Chen, X.; Zhou, T.; Yuan, H.; Chen, H. Structural stability, half-metallicity and magnetism of the CoFeMnSi/GaAs(0 0 1) interface. *Appl. Surf. Sci.* **2015**, *346*, 1–10. [CrossRef]

14. Knut, R.; Svedlindh, P.; Mryasov, O.; Gunnarsson, K.; Warnicke, P.; Arena, D.A.; Bjorck, M.; Dennison, A.J.C.; Sahoo, A.; Mukherjee, S.; et al. Interface characterization of Co2MnGe/Rh2CuSn Heusler multilayers. *Phys. Rev. B* **2013**, *88*, 134407. [CrossRef]

15. Rani, D.; Suresh, E.K.G.; Yadav, A.K.; Jha, S.N.; Varma, D.M.R.; Alam, A. Structural, electronic, magnetic, and transport properties of the equiatomic quaternary Heusler alloy CoRhMnGe: Theory and experiment. *Phys. Rev. B* **2017**, *96*, 184404. [CrossRef]

16. Yang, F.J.; Wei, C.; Chen, X.Q. Half-metallicity and anisotropic magnetoresistance of epitaxial Co2FeSi Heusler films. *Appl. Phys. Lett.* **2013**, *102*, 172403. [CrossRef]

17. Scheike, T.; Sukegawa, H.; Furubayashi, T.; Wen, Z.; Inomata, K.; Ohkubo, T.; Hono, K.; Mitani, S. Lattice-matched magnetic tunnel junctions using a Heusler alloy Co2FeAl and a cationdisorder spinel Mg-Al-O barrier. *Appl. Phys. Lett.* **2014**, *105*, 242407. [CrossRef]

18. Feng, Y.; Wu, B.; Yuan, H.; Chen, H. Structural, electronic and magnetic properties of Co2MnSi/Ag(100) interface. *J. Alloy. Compd.* **2015**, *623*, 29–35. [CrossRef]

19. Fetzer, R.; Wüstenberg, J.P.; Taira, T.; Uemura, T.; Yamamoto, M.; Aeschlimann, M.; Cinchetti, M. Structural, chemical, and electronic properties of the Co2MnSi(001)/MgO interface. *Phys. Rev. B* **2013**, *87*, 184418. [CrossRef]

20. Yamaguchi, T.; Moriya, R.; Oki, S.; Yamada, S.; Masubuchi, S.; Hamaya, K.; Machida, T. Spin injection into multilayer grapheme from highly spin-polarized Co2FeSi Heusler alloy. *Appl. Phys. Express* **2016**, *9*, 063006. [CrossRef]

21. Feng, Y.; Cui, Z.; Wei, M.; Wu, B. Spin-polarized quantum transport in Fe₄N based current-perpendicular-toplane spin valve. *Appl. Surf. Sci.* **2019**, *466*, 78–83. [CrossRef]
22. Lkhagvasuren, E.; Ouardi, S.; Fecher, G.H.; Auffermann, G.; Kreiner, G.; Schnelle, W.; Felser, C. Optimized thermoelectric performance of the n-type half-Heusler material TiNiSn by substitution and addition of Mn. *AIP Adv.* **2017**, *7*, 045010. [CrossRef]
23. Casper, F.; Graf, T.; Chadov, S.; Balke, B.; Felser, C. Half-Heusler compounds: Novel materials for energy and spintronic applications. *Semicond. Sci. Technol.* **2012**, *27*, 063001. [CrossRef]
24. Chen, S.; Ren, Z. Recent progress of half-Heusler for moderate temperature thermoelectric applications. *Mater. Today* **2013**, *16*, 387–395. [CrossRef]
25. Bos, J.W.G.; Downie, R.A. Half-Heusler thermoelectrics: A complex class of materials. *J. Phys. Condens. Matter* **2014**, *26*, 433201. [CrossRef] [PubMed]
26. Wang, X.L. Proposal for a new class of materials: Spin gapless semiconductors. *Phys. Rev. Lett.* **2008**, *100*, 156404. [CrossRef] [PubMed]
27. Ouardi, S.; Fecher, G.H.; Felser, C.; Kübler, J. Realization of Spin Gapless Semiconductors: The Heusler Compound Mn2CoAl. *Phys. Rev. Lett.* **2013**, *110*, 100401. [CrossRef] [PubMed]
28. Galanakis, I.; Özdoğan, K.; Şaolu, E.; Blügel, S. Conditions for spin-gapless semiconducting behavior in Mn2CoAl inverse Heusler compound. *J. Appl. Phys.* **2014**, *115*, 093908. [CrossRef]
29. Jamer, M.E.; Assaf, B.A.; Devakul, T.; Heiman, D. Magnetic and transport properties of Mn₂CoAl oriented films. *Appl. Phys. Lett.* **2013**, *103*, 2–7. [CrossRef]
30. Feng, Y.; Bo, W.; Yuan, H.; Kuang, A.; Chen, H. Magnetism and half-metallicity in bulk and (100) surface of Heusler alloy Ti2CoAl with Hg2CuTi-type structure. *J. Alloy. Compd.* **2013**, *557*, 202–208. [CrossRef]
31. Bainsla, L.; Mallick, A.I.; Raja, M.M.; Coelho, A.A.; Nigam, A.K.; Johnson, D.D.; Alam, A.; Suresh, K.G. Origin of spin gapless semiconductor behavior in CoFeCrGa: Theory and Experiment. *Phys. Rev. B* **2015**, *92*, 045201. [CrossRef]
32. Galanakis, I.; Özdoğan, K.; Şaşioğlu, E. Spin-filter and spin-gapless semiconductors: The case of Heusler compounds. *AIP Adv.* **2016**, *6*, 055606. [CrossRef]
33. Gao, G.Y.; Yao, K.L. Antiferromagnetic half-metals, gapless half-metals, and spin gapless semiconductors: The D03-type Heusler alloys. *Appl. Phys. Lett.* **2013**, *103*, 232409. [CrossRef]
34. Xu, G.Z.; Liu, E.K.; Du, Y.; Li, G.J.; Liu, G.D.; Wang, W.H.; Wu, G.H. A new spin gapless semiconductors family: Quaternary Heusler compounds. *EPL* **2013**, *102*, 17007. [CrossRef]
35. Bainsla, L.; Suresh, K.G. Equiatomic quaternary Heusler alloys: A material perspective for spintronic applications. *Appl. Phys. Rev.* **2016**, *3*, 031101. [CrossRef]
36. Perdew, J.P.; Burke, K.; Ernzerhof, M. Generalized gradient approximation made simple. *Phys. Rev. Lett.* **1996**, *77*, 3865. [CrossRef] [PubMed]
37. Blöchl, P.E. Projector augmented-wave method. *Phys. Rev. B* **1994**, *50*, 17953. [CrossRef]

applied
sciences

MDPI

Article

Electronic and Magnetic Properties of Bulk and Monolayer CrSi$_2$: A First-Principle Study

Shaobo Chen [1,*] , Ying Chen [1], Wanjun Yan [1], Shiyun Zhou [1], Xinmao Qin [1], Wen Xiong [2] and Li Liu [3]

[1]   College of Electronic and Information Engineering, Anshun University, Anshun 561000, China;
      ychenjz@163.com (Y.C.); yanwanjun7817@163.com (W.Y.); s.y.zhou@163.com (S.Z.);
      qxm200711@126.com (X.Q.)
[2]   Department of Physics and Institute of Condensed Matter Physics, Chongqing University,
      Chongqing 400000, China; wenxiong@cqu.edu.cn
[3]   Yichang No.1 Senior High School, Yichang 443000, China; ycyzliuli@163.com
*   Correspondence: chenshaobo@asu.edu.cn; Tel.: +86-0851-3221-4631

Received: 11 September 2018; Accepted: 1 October 2018; Published: 11 October 2018

Abstract: We investigated the electronic and magnetic properties of bulk and monolayer CrSi$_2$ using first-principle methods based on spin-polarized density functional theory. The phonon dispersion, electronic structures, and magnetism of bulk and monolayer CrSi$_2$ were scientifically studied. Calculated phonon dispersion curves indicated that both bulk and monolayer CrSi$_2$ were structurally stable. Our calculations revealed that bulk CrSi$_2$ was an indirect gap nonmagnetic semiconductor, with 0.376 eV band gap. However, monolayer CrSi$_2$ had metallic and ferromagnetic (FM) characters. Both surface and confinement effects played an important role in the metallic behavior of monolayer CrSi$_2$. In addition, we also calculated the magnetic moment of unit cell of 2D multilayer CrSi$_2$ nanosheets with different layers. The results showed that magnetism of CrSi$_2$ nanosheets was attributed to band energy between layers, quantum size, and surface effects.

Keywords: electronic property; magnetism; bulk CrSi$_2$; monolayer CrSi$_2$; first-principle

1. Introduction

Since graphene [1], which is widely used in the fields of materials, electronics, physics, chemistry, energy resources, biomedicines, etc., was discovered by Andre Geim and Konstantin Novoselov, two-dimensional (2D) layered materials have triggered extensive interest owing to their unique physical properties [2]. In the past decades, two-dimensional (2D) materials such as silicene, h-BN, layered transition metal dichalcogenides (TMD), and monolayer transition metal silicides (TMSi$_2$) have been widely studied [3–11]. Contrasting with bulk materials, low-dimensional materials with unusual physical properties are important for potential applications in spintronics [12–15], magnetic storage [16,17], and molecular scale electronic devices [18,19], etc. Two-dimensional (2D) materials present extensive novel properties due to quantum size effects [20–23]. The properties of materials strongly depend on the crystal structure. Thus, we can change structure phases to tune the properties of materials. Previous investigations have proved that controlling the crystal structure and thickness of materials can tune magnetic moment [20], transform metal to semimetal or semiconductor transition [21], and phase segregation [22], as well as alter electronic properties [23]. Among transition metal silicide, CrSi$_2$ received numerous attention due to important applications in Si-based device technology [24–26]. Previous literature concludes that bulk CrSi$_2$ is an indirect semiconductor. Nevertheless, to the best of our knowledge, few studies [27,28] on their magnetic properties have been reported. In recent years, studies have demonstrated that depending on the compositions, 2D monolayer transition metal silicides (TMSi$_2$) are found to be either magnetic or nonmagnetic [9–11]. Among TMSi$_2$ monolayers, CrSi$_2$ sheet is found to be ferromagnetic [9–11,29], thus it may become an important magnetic nanomaterial in spintronics. Theoretically, both Dzade et al. [29] and Bui et al. [30] have employed the quantum ESPRESSO package to investigate silicene and transition metal-based materials. Interestingly, they get different even conflicting results, i.e., Dzade deduced that two-dimensional CrSi$_2$ is ferromagnetic, whilst Bui deduced that planar CrSi$_2$ favors anti-ferromagnetism.

The development of spintronics and magnetic storage urgently needs synthesized novel 2D magnetic materials. Recently, functionalization of nonmagnetic monolayer materials has been a major way to induce magnetism [31–33]. Inspired by the synthesis of silicon, both theoretical [29,30] and experimental [34] researchers have studied the properties of transition metal silicides layers. There is a big obstacle in synthesizing truly two-dimensional nanomaterials and it is because its structural stability depends on temperature sensitively. Naturally, more 2D magnetic materials are required to meet the demand for the rapid development of spintronics and magnetic storage. In this paper, first-principle calculations are employed to probe how dimension and size affect the electronic structure and magnetism of both bulk and monolayer CrSi$_2$. The phonon dispersion curve, band structures with spin state, total and partial density of states (DOS), spin density of bulk and monolayer CrSi$_2$ system, and magnetic moment of per unit cell of multilayer CrSi$_2$ nanosheets varying with different layers, are systematically investigated. These results suggest monolayer CrSi$_2$ may have potential applications in exploiting molecular scale electronic devices.

2. Material and Methods

Our calculations were performed using spin-polarized density functional theory (DFT) in the generalized gradient approximation (GGA) [35,36], with the Pedew-Burke-Ernzerhof (PBE) function for exchange-correlation potential, which were implemented in the Cambridge Sequential Total Energy Package (CASTEP) [37]. Projector augmented-wave (PAW) potentials [38] were employed to illustrate electron–ion interactions. The convergence criterion of total energy was set to be 10^{-6} eV, and energy cutoff of 310 eV was adopted for the expansion of plane waves after our test. The Monkhorst-Pack [39] k-point grids of $6 \times 6 \times 6$, $6 \times 6 \times 1$ were applied for the Brillouin-zone (BZ) integration in bulk, and monolayer CrSi$_2$ computation, respectively. For monolayer CrSi$_2$, vacuum-slabs of 15Å were used to avoid interactions between adjacent atom layers. CrSi$_2$ has a hexagonal structure (C40) with nonsymmorphic space group D_6^4-P6$_2$22 [24,25], containing no primitive translations which interchange individual CrSi$_2$ layers. The lattice constants were a = 4.431 and c = 6.364 Å [24,25]. The lattice constants and atomic positions were fully relaxed until the force on each atom was less than 0.03 eV/Å. Monolayer CrSi$_2$ has a graphene-like honeycomb structure, which can be formed by a micromechanical cleavage technique due to weak van der Waals (vdW) forces between those layers and strong covalent bonding intralayer [40]. Top and side views of monolayer CrSi$_2$ after geometry optimization are depicted in Figure 1. According to chemical formula, per unit cell is constructed by one Cr atom and two Si atoms because in every intralayer, one Cr atom is in the center of each hexagonal hole of silicene lattice, leading to a 1:2 ratio between Cr and Si.

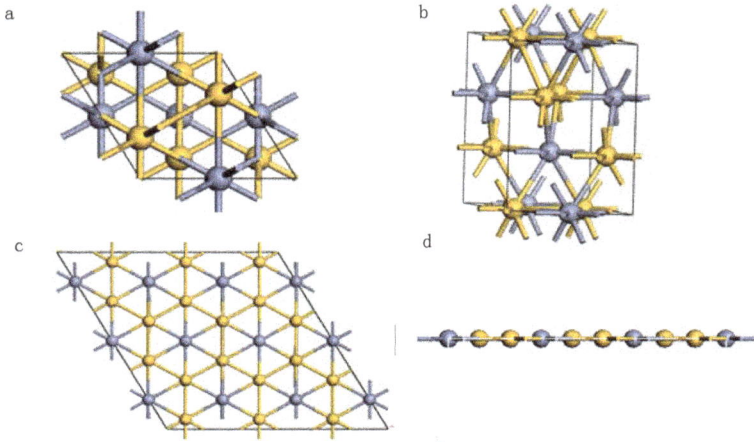

Figure 1. Top and side views of bulk and monolayer CrSi$_2$ after geometry optimization. Figure 1**a,b** describe top view and side view of the bulk CrSi$_2$ crystalline structure, Figure 1**c,d** depict top view and side view of the monolayer CrSi$_2$ crystalline structure, respectively. Yellow balls and blue balls represent Si and Cr atoms.

3. Results and Discussion

3.1. Phonon Dispersion Curve

It is necessary to check the structural stability of materials before calculation. Although the structures of bulk and monolayer CrSi$_2$ have been optimized, the phase stability of these structures remains uncertain. Phonon dispersion spectrum analysis is a valid tool to confirm the structural stability. If all the phonon frequencies on the k-points in the Brillouin zone are positive, the structure is stable at absolute zero of temperature. Otherwise, the structure is unstable at absolute zero of temperature [41]. To check the structural stability of bulk and monolayer CrSi$_2$, we accurately calculated phonon dispersion curves along the high symmetry directions in the Brillouin zone. As shown in Figure 2a for bulk CrSi$_2$, and b for monolayer CrSi$_2$, no imaginary vibration frequency appears for bulk and monolayer CrSi$_2$, indicating that both structures of bulk and monolayer CrSi$_2$ are stable at ground state in accordance with-Ref. [41].

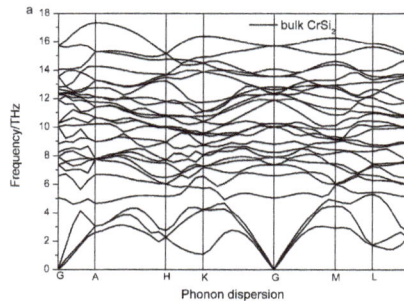

Figure 2. *Cont.*

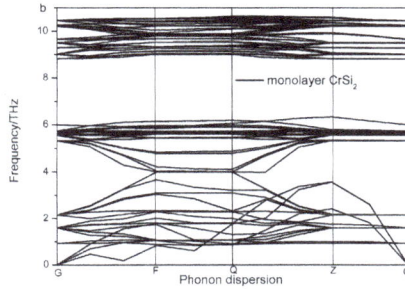

Figure 2. Phonon dispersion curves along the high-symmetry directions in the Brillouin zone of (**a**) bulk $CrSi_2$, (**b**) monolayer $CrSi_2$.

3.2. Electronic Structure

We calculated the magnetic moment of unit cell, local magnetic moment of Cr and Si atom, total energy, band length between Cr and Si atoms, band gap and lattice parameters, listed in Table 1. The results were satisfying, compared with those values calculated in References [9,24,25,29,30]. One can see that bulk $CrSi_2$ is an indirect gap semiconductor, whereas, monolayer $CrSi_2$ has metallic character. It can also be seen that there is a big difference between bulk and monolayer compounds in the magnetic moment, i.e., monolayer $CrSi_2$ unit cell has an obvious magnetic moment~3.68 μ_B, and the local magnetic moment of every Cr and Si atom were 4.11 and −0.21 μ_B, respectively. In contrast, for the bulk $CrSi_2$ system, every Cr and Si atom hardly had any magnetic moment. These results indicated that whilst bulk $CrSi_2$ was diamagnetic, monolayer $CrSi_2$ system was ferromagnetic (FM), consistent with the conclusions of References [9–11,29]. However, it conflicts with the results of Reference [30]. Unfortunately, until now, there is no available experimental evidence to validate the contradicting theoretical results. Potentially, this research may inspire more experimenters to study these two-dimensional systems.

Table 1. Magnetic moment and structure of bulk and monolayer $CrSi_2$.

	Magnetic Moment of Unit Cell (μ_B)	Local Magnetic Moment of Cr Atom (μ_B)	Local Magnetic Moment of Si Atom (μ_B)	Total Energy of System (ev)	Band Length of cr-si of Intralayer (å)	Band Gap (eV)	Lattice Parameter (Å)
bulk $CrSi_2$	4×10^{-4}	0	0	−8050.32	2.47, 2.52,2.55	0.376	a = 4.4276 c = 6.3681
	0 [c]	0 [c]	0 [c]	–	2.47 [a], 2.55 [a], 3.06 [a]	0.35 [a], 0.21 [d]	a = 4.42 [a], 4.43 [d] c = 6.349 [a], 6.36 [d]
monolayer $CrSi_2$	3.68	4.11	−0.21	−24118.24	2.55	0	a = 4.4276 c = 15
	3.6 [b]	4.15 [c]	–	–	2.56 [b]	0 [c]	a = 3.93968 [e] c = 16.49899 [e]

[a] Reference [25]. [b] Reference [29]. [c] Reference [30]. [d] Reference [24]. [e] Reference [9].

To reveal the origin of metallicity and magnetism, band structure and total and partial density of states (DOS) were systematically studied. As shown in Figure 3, the band structures with up and down spin of bulk and monolayer $CrSi_2$ are calculated. The results show that bulk $CrSi_2$ is an indirect gap semiconductor with a band gap of 0.376 eV, which is in good accordance with Ref. [24,25], and monolayer $CrSi_2$ is metallic being in good agreement with our previous results [10,11]. In the bulk $CrSi_2$ system, the spin-up and spin-down states were completely symmetric, which indicated that bulk $CrSi_2$ was a nonmagnetic semiconductor. However, for the monolayer $CrSi_2$ system, the spin-up and spin-down states were in complete asymmetry and both spin-up and spin-down states go across the Fermi level, which manifested that monolayer $CrSi_2$ was both magnetic and metallic. All these results were in good agreement with the analysis of Table 1. It has been confirmed that the energy band structure of TMDs are greatly affected by the crystal structure [20]. Huang et al. [42] have explored the origin of the high metallicity on $MoSi_2$ nanofilms in detail. We can elucidate the physics mechanism of why bulk $CrSi_2$ is a semiconductor, whilst monolayer $CrSi_2$ is metallic using Huang's theory. Both surface and confinement effects contribute to the high sensitivity of the metallicity on nanofilms type, explaining the reason why monolayer has a metallic character.

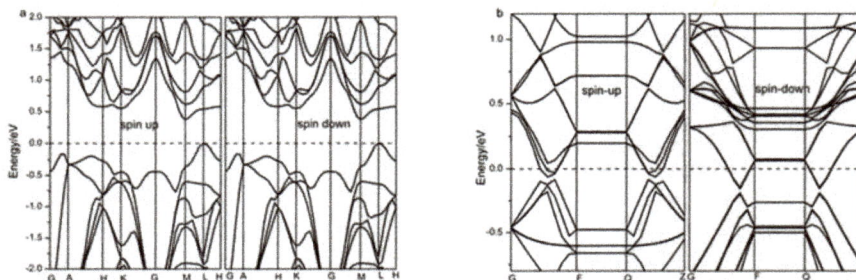

Figure 3. Band structure with spin-up and spin-down of (**a**) bulk and (**b**) monolayer $CrSi_2$.

To further investigate the physical mechanism of magnetism, which may be dependent on the dimension of materials, we calculated the total density of states (DOS) and partial density of states (PDOS) of bulk and monolayer $CrSi_2$ systems, as depicted in Figure 4. Total DOS and PDOS of the bulk $CrSi_2$ system were fully symmetric, indicating that bulk $CrSi_2$ system cannot have a magnetic characteristic, in accordance with Figure 3a. In contrast, for the monolayer $CrSi_2$ system, both total DOS and PDOS of monolayer $CrSi_2$ system were asymmetric, manifesting that monolayer $CrSi_2$ system possesses a magnetic characteristic, in good agreement with Figure 3b. The degree of dissymmetry in PDOS of the Cr atom in monolayer $CrSi_2$ system was greater than the Si atom's, which is the reason why the Cr atom has a larger local magnetic moment as depicted in Table 1. In addition, the total density of state near Fermi level of the bulk $CrSi_2$ system mainly consists of Cr-3d orbital electron. The total density of state near Fermi level of the monolayer $CrSi_2$ system is mainly made up of Cr-3d orbital electron, with Cr-3p and Si-3p orbital electrons making limited contribution to the total density of state of the system. Moreover, the results also indicated that total magnetic moment (3.68 µB) arose mainly from the spin-up Cr-3d states. Han [9] has investigated the origin of magnetic behavior in monolayer $FeSi_2$ and $CoSi_2$ by orbital coupling of atoms. The stronger orbital coupling between atoms may account for the quench of magnetism of the atom. It can be seen from Figure 4b that no noticeable coupling between p orbital of Si atom and d orbital of Cr atom is found around the Fermi level, which indicates that monolayer $CrSi_2$ has magnetic behavior.

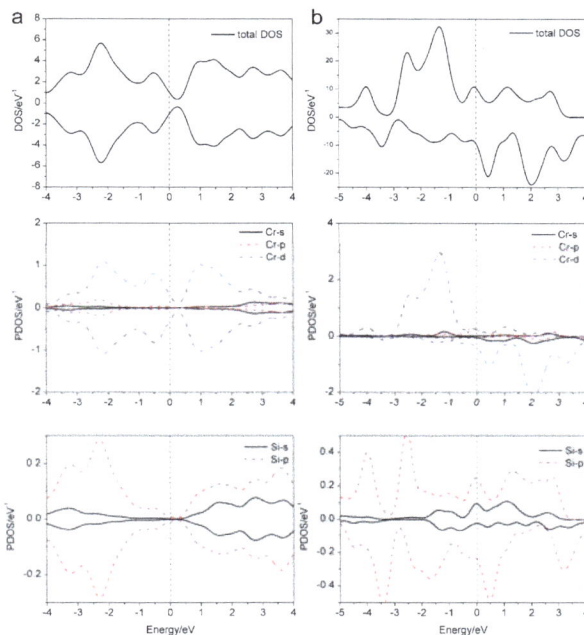

Figure 4. Total and partial density of states (DOS) varies from the energy for (**a**) bulk and (**b**) monolayer $CrSi_2$.

3.3. Magnetic Properties

To understand the origin of magnetism, which may be dependent on the dimension of the material, we further investigated the spin density of bulk and monolayer $CrSi_2$ systems. As shown in Figure 5, the spin density isosurface plots of bulk and monolayer $CrSi_2$ on top view (001) are particularly calculated. For the bulk $CrSi_2$ system, spin density near Cr and Si atoms was close to zero, which agreed with the calculations of local magnetism moment of Cr and Si atoms (0 μ_B) in Table 1. Nevertheless, for the monolayer $CrSi_2$ system, the numerical values of spin density near Cr atoms were very noticeable, which was much larger than those of spin density near Si atoms. It indicated that the behavior of magnetism in the monolayer was mainly contributed by the magnetic property of Cr atoms. Through careful analysis, we found that electron transfer from one Cr atom to one Si was equal in both the bulk and monolayer $CrSi_2$ systems. The magnetic behavior has discrepancy in different dimension structures, which can be interpreted considering the charge transfer model [43] and Hund's rules. The valence electron configurations of Cr and Si atoms are $3d^5 4s^1$ and $3s^2 3p^2$, respectively. In the bulk structure, every Cr ($3d^5 4s^1$) atom transfers one 4s electron and one 3d electron to adjacent two Si ($3s^2 3p^2$) atoms. Then, the Si atom whose electron configuration is $3s^2 3p^3$ captures one electron to form a stable close-shell electronic structure, and thus has zero spin. The electron configuration of Cr atom is $3d^4$, which is an unstable electronic structure according to the octet rule. Owing to the van der Waals (vdW) force and chemical bonds energy between layers, valence electrons of Cr atom are antiparallel, as depicted in Figure 6a, leaving neither unpaired electrons nor net spin, which demonstrates that the Cr atom has no magnetic moment in the bulk $CrSi_2$ system. This is slightly different from spin density, as depicted in Figure 5a, because it does not consider crystal field splitting. In the monolayer structure, valence electrons of Cr atom are parallel, as depicted in Figure 6b, which has the lowest energy due to the absence of the van der Waals (vdW) force and chemical bonds between layers, as well as the decline of chemical bonds energy in intralayer (i.e., the bond lengths increase, see Table 1), leaving unpaired electrons and net spin. This demonstrates that the Cr atom has magnetic moment in the monolayer $CrSi_2$ system. Compared with the bulk material, electrons in the monolayer case favored occupying different orbits and having parallel spins, resulting in the monolayer case having less unfavorable Coulomb repulsion and lower energy. It was consistent with the Hund's rules, that electrons always take precedence of different orbits and have the same spin direction occupying the equivalent orbital.

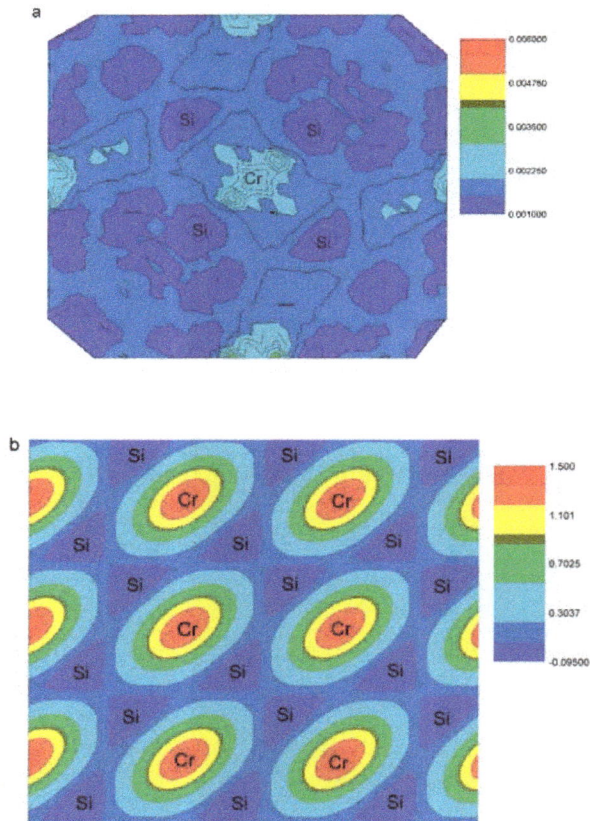

Figure 5. Spin density iso-surface plots of (**a**) bulk and (**b**) monolayer $CrSi_2$ on top view (001). The iso-surface level is set as $0.003e/\text{Å}^3$.

Owing to quantum size and surface effects, two-dimensional (2D) materials may present extensive novel physical and chemical properties, when downsizing from three dimensions to two dimensions or one dimension [20–23,44–47]. To further investigate the interrelationship between thickness and magnetism in $CrSi_2$, we also calculated magnetic moment of unit cell of 2D multilayer nanosheets with different layers. The results depicted in Figure 7 show that the magnetic moment sharply decreases with the increase in the numbers of layers (especially, the magnetic moment decreases greatest when the number of layers increases from one layer to two layers). As the layers increase, the decrease of magnetic moments occurs in $CrSi_2$ nanosheets. Considering weak van der Waals force and strong chemical bonds between layers, quantum size [20–23,46] and surface effects [44] occur when downsizing from bilayers to monolayer in $CrSi_2$ nanosheets. We deduced that the band energy between layers, as well as quantum size and surface effects play an important role in magnetism of materials. Magnetic response of 2D materials can be tuned by controlling the thinness of thin films, which is an advantageous application in magnetic materials.

Figure 6. Valence electron diagrams of Cr atom in (**a**) bulk and (**b**) monolayer $CrSi_2$ systems.

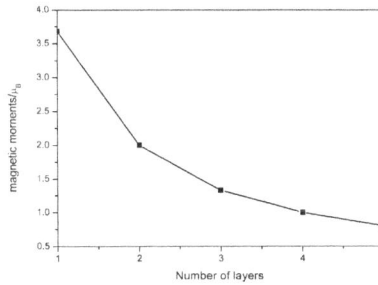

Figure 7. The magnetic moment of unit cell of multilayer $CrSi_2$ nanosheets varies with different layers.

4. Conclusions

Electronic and magnetic properties of $CrSi_2$ were calculated using the first-principle methods based on density functional theory. The phonon dispersion curve, band structures with spin state, total and partial density of states (DOS), and spin density of bulk and monolayer $CrSi_2$ systems were systematically investigated. Both bulk and monolayer $CrSi_2$ were structurally stable at ground state. The results showed that the bulk $CrSi_2$ system is an indirect gap non-magnetic semiconductor with a band gap of 0.376 eV, whilst the monolayer $CrSi_2$ sheets were metallic and ferromagnetic (FM). Compared with previous literature, our results were consistent with Dzade's results (ferromagnetism) and inconsistent with Bui's results (anti-ferromagnetism). We explain the reason why monolayer $CrSi_2$ had metallic behavior using Huang's theory, where surface and confinement effects play an important role in the metallic behavior of monolayer $CrSi_2$. Further analysis showed that total DOS and PDOS of the bulk $CrSi_2$ system were fully symmetric, and those of the monolayer $CrSi_2$ system were asymmetric, which may reveal the physical mechanism of magnetism for bulk and $CrSi_2$ nanosheets. In addition, we also elucidated the origin of magnetism considering the charge transfer model and Hund's rules, where magnetic moment of unit cell of 2D multilayer nanosheets with different layers was calculated. The results showed that the magnetism of materials is attributed to band energy between layers, as well as quantum size and surface effects. We also expect that our calculations may provide some helpful insight into further experimental investigations, and they show promise in device applications based on 2D $CrSi_2$ nanosheets.

Author Contributions: Conceptualization, S.C.; Methodology, Y.C. and X.Q.; Software, S.C. and Y.C.; Formal Analysis, W.Y., S.Z., and W.X.; Data Curation, S.C.; Writing-Original Draft Preparation, S.C.; Writing-Review & Editing, S.C., L.L., and W.X.; Supervision, Y.C.

Funding: This research was funded by [Key Projects of the Tripartite Foundation of Guizhou Science and Technology Department] grant number [[2015]7696], and by [Guizhou College Student Innovation Entrepreneurship Training Program] grant number [201710667017], and by [Major Projects for Creative Research Groups of Guizhou Province of China] grant number [[2016]048], and by [Innovation Team of Anshun University] grant number [2015PT02], and by [Natural Science Foundation of Science and Technology Department of Guizhou Province of China] grant number [[2010]2001].

Acknowledgments: We are really grateful to the cloud computing platform at Guizhou University and the CNROCK HOLE BLACKHOLE high density computing platform for computing support.

Conflicts of Interest: The authors declare no conflict of interest.

References

1. Novoselov, K.S.; Geim, A.K.; Morozov, S.V.; Jiang, D.; Zhang, Y.; Dubonos, S.V.; Grigorieva, I.V.; Firsov, A.A. Electric field effect in atomically thin carbon films. *Science* **2004**, *306*, 666–669. [CrossRef] [PubMed]
2. Xu, M.S.; Liang, T.; Shi, M.M.; Chen, H.Z. Graphene-like two-dimensional materials. *Chem. Rev.* **2013**, *113*, 3766–3798. [CrossRef] [PubMed]
3. Lew Yan Voon, L.C.; Guzmán-Verri, G.G. Is silicone the next graphence. *MRS Bull.* **2014**, *39*, 366–373. [CrossRef]
4. Kara, A.; Enriquez, H.; Seitsonen, A.P.; Lew Yan Voon, L.C.; Vizzini, S.; Aufray, B.; Oughaddou, H. A review on silicone-new candidate for electronics. *Sur. Sci. Rep.* **2012**, *67*, 1–18. [CrossRef]
5. Golberg, D.; Bando, Y.; Huang, Y.; Terao, T.; Mitome, M.; Tang, C.; Zhi, C. Boron nitride nanotubes and nanosheets. *ACS Nano* **2010**, *4*, 2979–2993. [CrossRef] [PubMed]
6. Pakdel, A.; Zhi, C.; Bando, Y.; Golberg, D. Low-dimensional boron nitride nanomaterials. *Mat. Today* **2012**, *15*, 256–265. [CrossRef]
7. Chhowalla, M.; Shin, H.S.; Eda, G.; Li, L.J.; Loh, K.P.; Zhang, H. The chemistry of two-dimensional layered transition metal dichalcogenidenanosheets. *Nat. Chem.* **2013**, *5*, 263–275. [CrossRef] [PubMed]
8. Butler, S.Z.; Hollen, S.M.; Cao, L.; Cui, Y.; Gupta, J.A. Progress, challenges, and opportunities in two-dimensional materials beyond grapheme. *ACS Nano* **2013**, *7*, 2898–2926. [CrossRef] [PubMed]
9. Han, N.N.; Liu, H.S.; Zhao, J.J. Novel magnetic monolayers of transition metal silicide. *J. Superconduct. Nov. Magn.* **2015**, *28*, 1755–1758. [CrossRef]
10. Chen, S.B.; Chen, Y.; Yan, W.J.; Zhou, S.Y.; Xiong, W.; Yao, X.X.; Qin, X.M. Magnetism and optical property of Mn-doped monolayer $CrSi_2$ by first-principle study. *J. Superconduct. Nov. Magn.* **2017**. [CrossRef]
11. Chen, S.B.; Zhou, S.Y.; Yan, W.J.; Chen, Y.; Qin, X.M.; Xiong, W. Effect of Fe and Ti Substitution Doping on Magnetic Property of Monolayer $CrSi_2$: A first-principle investigation. *J. Superconduct. Nov. Magn.* **2018**. [CrossRef]
12. Han, W.; Kawakami, R.K.; Gmitra, M.; Fabian, J. Graphene spintronics. *Nat. Nanotechnol.* **2014**, *59*, 794–807. [CrossRef] [PubMed]
13. Maassen, J.; Ji, W.; Guo, H. Graphene spintronics: The role of ferromagnetic electrodes. *Nano Lett.* **2011**, *11*, 151–155. [CrossRef] [PubMed]
14. Fuh, H.R.; Chang, K.W.; Hung, S.H.; Jeng, H.T. Two-dimensional magnetic semiconductors based on transition-metal dichalcogenides VX_2 (X=S, Se, Te) and similar layered compounds VI_2 and $Co(OH)_2$. *IEEE Magn. Lett.* **2017**, *8*, 3101405. [CrossRef]
15. Das Sarma, S.; Adam, S.; Hwang, E.H.; Rossi, E. Electronic transport in two dimensional graphene. *Rev. Mod. Phys.* **2011**, *83*, 407–470. [CrossRef]
16. Parkin, S.; Yang, S.H. Memory on the racetrack. *Nat. Nanotechnol.* **2015**, *10*, 195–198. [CrossRef] [PubMed]
17. Parkin, S.S.P.; Hayashi, M.; Thomas, L. Magnetic domain-wall racetrack memory. *Science* **2008**, *320*, 190–194. [CrossRef] [PubMed]
18. Eda, G.; Fujita, T.; Yamaguchi, H.; Voiry, D.; Chen M Chhowalla, M. Coherent atomic and electronic heterostructures of single-layer MoS_2. *ACS Nano* **2012**, *6*, 7311–7317. [CrossRef] [PubMed]
19. Ci, L.; Song, L.; Jin, C.; Jariwala, D.; Wu, D.; Li, Y. Atomic layers of hybridized boron nitride and graphene domains. *Nat. Mater.* **2010**, *9*, 430–435. [CrossRef] [PubMed]
20. Zhang, H.; Liu, L.M.; Lau, W.M. Dimension-dependent phase transition and magnetic properties of VS2. *J. Mater. Chem. A* **2013**, *1*, 10821–10828. [CrossRef]
21. Abdul Wasey, A.H.M.; Soubhik, C.; Das, G.P. Quantum size effects in layered VX_2 (X=S, Se) materials: Manifestation of metal to semimetal or semiconductor transition. *J. Appl. Phys.* **2015**, *117*, 064313. [CrossRef]
22. Tan, C.L.; Sun, D.; Tian, X.H.; Huang, Y.W. First-principles investigation of phase stability, electronic structure and optical properties of MgZnO monolayer. *Materials* **2016**, *9*, 877. [CrossRef] [PubMed]
23. Wang, W.D.; Bai, L.W.; Yang, C.G.; Fan, K.Q.; Xie, Y.; Lin, M.L. The electronic properties of O-doped pure and sulfur vacancy-defect monolayer WS_2: A first-principles study. *Materials* **2018**, *11*, 218. [CrossRef] [PubMed]
24. Krijn, M.P.C.M.; Eppenga, R. First-principles electronic structure and optical properties of $CrSi_2$. *Phys. Rev. B* **1991**, *44*, 9042–9044. [CrossRef]
25. Mattheiss, L.F. Electronic structure of $CrSi_2$ and related refractory disilicides. *Phys. Rev. B* **1991**, *43*, 12549–12555. [CrossRef]

26. Dasgupta, T.; Etourneau, J.; Chevalier, B.; Matar, S.F.; Umarji, A.M. Structural, thermal, and electrical properties of CrSi$_2$. *J. Appl. Phys.* **2008**, *103*, 113516. [CrossRef]

27. Singh, D.J.; Parker, D. Itinerant magnetism in doped semiconducting β-FeSi2 and CrSi$_2$. *Sci. Rep.* **2013**, *3*, 3517. [CrossRef] [PubMed]

28. Parker, D.; Singh, D.J. Very heavily electron-doped CrSi$_2$ as a high performance high-temperature thermoelectric material. *New J. Phys.* **2012**, *14*, 033045. [CrossRef]

29. Dzade N, Y.; Obodo, K.O.; Adjokatse, S.K. Silicene and transition metal based materials: Prediction of a two dimensional piezomagnet. *J. Phys. Condens. Matter.* **2010**, *22*, 375502–375509. [CrossRef] [PubMed]

30. Viet Q, B.; Pham, T.T.; Nguyen, H.V.S.; Le, H.M. Transition metal (Fe and Cr) adsorptions on buckled and planar silicene monolayers: A density functional theory investigation. *J. Phys. Chem. C* **2013**, *117*, 23364–23371.

31. Wang, X.Q.; Li, H.D.; Wang, J.T. Induced ferromagnetism in one-side semihydrogenated silicene and germanene. *Phys. Chem. Chem. Phys.* **2012**, *14*, 3031–3036. [CrossRef] [PubMed]

32. Zhang, C.W.; Yan, S.S. First-principles study of ferromagnetism in two-dimensional silicene with Hydrogenation. *J. Phys. Chem. C* **2012**, *116*, 4163–4166. [CrossRef]

33. Kaloni, T.P.; Gangopadhyay, S.; Singh, N.; Jones, B. Electronic properties of Mn-decorated silicene on hexagonal boron nitride. *Phys. Rev. B* **2013**, *88*, 235418. [CrossRef]

34. Zhu, H.N.; Gao, K.Y.; Liu, B.X. Formation of n-type CrSi$_2$ semiconductor layers on Si by high-current Cr ion implantation. *J. Phys. D Appl. Phys.* **2000**, *33*, L49–L52. [CrossRef]

35. Perdew, J.P.; Burke, K.; Ernzerhof, M. Generalized gradient approximation made simple. *Phys. Rev. Lett.* **1996**, *77*, 3865–3868. [CrossRef] [PubMed]

36. Payne, M.C.; Teter, M.P.; Allan, D.C.; Arias TA Joannopoulos, J.D. Iterative minimization techniques for ab initio total-energy calculations: Molecular dynamics and conjugate gradients. *Rev. Mod. Phys.* **1992**, *64*, 1064–1096. [CrossRef]

37. Clark, S.J. First principles methods using CASTEP. *Z. Kristall.* **2005**, *220*, 567–570. [CrossRef]

38. Kresse, G.; Joubert, D. Self-interaction correction to density-functional approximations for many-electron systems. *Phys. Rev. B Condens. Matter Mater. Phys.* **1999**, *59*, 1758–1775. [CrossRef]

39. Monkhorst, H.J.; Pack, J.D. Special points for Brillouin-zone integrations. *Phys. Rev. B* **1976**, *13*, 5188–5192. [CrossRef]

40. Zeng, Z.Y.; Yin, Z.Y.; Huang, X.; Li, H.; He, Q.Y.; Lu, G.; Boey, F.; Zhang, H. Single-layer semiconducting nanosheets: High-yield preparation and device fabrication. *Angew. Chem. Int. Ed.* **2011**, *50*, 11093–11097. [CrossRef] [PubMed]

41. Hermet, P.; Khalil, M.; Viennois, R.; Beaudhuin, M.; Bourgogne, D.; Ravot, D. Revisited phonon assignment and electromechanical properties of chromium disilicide. *RSC Adv.* **2015**, *5*, 19106–19116. [CrossRef]

42. Huang L, R.; Rondinelli, J.M. Stable MoSi$_2$ nanofilms with controllable and high metallicity. *Phys. Rev. Mater.* **2017**, *1*, 063001-1–063001-6. [CrossRef]

43. Chen, Q.; Wang, J.L. Structural, electronic, and magnetic properties of TMZn$_{11}$O$_{12}$ and TM$_2$Zn$_{10}$O$_{12}$ clusters (TM = Sc, Ti, V, Cr, Mn, Fe, Co, Ni, and Cu). *Chem. Phys. Lett.* **2009**, *474*, 336–341. [CrossRef]

44. Li, H.; Qi, X.; Wu, J.; Zeng, Z.; Wei, J.; Zhang, H. Investigation of MoS$_2$ and graphene nanosheets by magnetic force microscopy. *ACS Nano* **2013**, *7*, 2842–2849. [CrossRef] [PubMed]

45. Tongay, S.; Varnoosfaderani, S.S.; Appleton, B.R.; Wu, J.Q.; Hebard, A.F. Magnetic properties of MoS$_2$: Existence of ferromagnetism. *Appl. Phys. Lett.* **2012**, *101*, 123105. [CrossRef]

46. Li, X.M.; Tao, L.; Chen, Z.F.; Fang, H.; Li, X.S.; Wang, X.R.; Xu, J.B.; Zhu, H.W. Graphene and related two-dimensional materials: Structure-property relationships for electronics and optoelectronics. *Appl. Phys. Rev.* **2017**, *4*, 021306. [CrossRef]

47. Li, Z.Q.; Chen, F. Ion beam modification of two-dimensional materials: Characterization, properties, and applications. *Appl. Phys. Rev.* **2017**, *4*, 011103. [CrossRef]

applied
sciences

MDPI

Article

Strain Control of the Tunable Physical Nature of a Newly Designed Quaternary Spintronic Heusler Compound ScFeRhP

Zongbin Chen [1], Habib Rozale [2], Yongchun Gao [1,*] and Heju Xu [1]

[1]   Department of Physics, College of Science, North China University of Science and Technology, Tangshan 063210, China; chen.12345@126.com (Z.C.); xuheju@126.com (H.X.)
[2]   Condensed Matter and Sustainable Development Laboratory, MDD Department, Faculty of Science, University of Sidi-Bel-Abbes, Sidi-Bel-Abbes 22000, Algeria; hrozale@yahoo.fr
*   Correspondence: gaoyc1963@126.com or gaoyc1963@ncst.edu.cn

Received: 19 August 2018; Accepted: 4 September 2018; Published: 7 September 2018

Abstract: Recently, an increasing number of rare-earth-based equiatomic quaternary compounds have been reported as promising novel spintronic materials. The rare-earth-based equiatomic quaternary compounds can be magnetic semiconductors (MSs), spin-gapless semiconductors (SGSs), and half-metals (HMs). Using first-principle calculations, we investigated the crystal structure, density of states, band structure, and magnetic properties of a new rare-earth-based equiatomic quaternary Heusler (EQH) compound, ScFeRhP. The results demonstrated that ScFeRhP is a HM at its equilibrium lattice constant, with a total magnetic moment per unit cell of 1 μ_B. Furthermore, upon introduction of a uniform strain, the physical state of this compound changes with the following transitions: non-magnetic-semiconductor-(NMS) $\rightarrow$ MS $\rightarrow$ SGS $\rightarrow$ HM $\rightarrow$ metal. We believe that these results will inspire further studies on other rare-earth-based EQH compounds for spintronic applications.

Keywords: quaternary Heusler compound; first-principle calculations; physical nature

---

1. Introduction

Since the first report on half-metal (HM) by Groot et al. [1], various Heusler compounds have been verified by theoretical approaches and experiments to be HM materials. The HM Heusler alloys attract significant interest owing to their novel physical properties [2–6].

Typically, Heusler-based HMs can be divided into three types: half-Heusler HMs [3], full-Heusler HMs [4], and quaternary Heusler HMs [5]. The quaternary Heusler alloys (XMYZ) can be regarded as a combination of two full-Heusler alloys: X_2YZ and M_2YZ (X, Y, M are individual 3d or 4d transition elements, while Z is an atom of the main group). Many quaternary Heusler HMs have been investigated by first-principle approaches. For example, Han et al. [6] have investigated a novel equiatomic quaternary Heusler (EQH) alloy YRhTiGe, concluding that it is an HM with a ferromagnetic ground state. Moreover, they studied its mechanical anisotropy, as well as dependence on the direction of shear modulus and Young's modulus in detail. New quaternary HMs, FeRuCrP and FeRhCrP [5], have been studied by Ma et al. in 2017; strain has been introduced to investigate its effect on the HM states.

In recent years, Wang [7] has theoretically predicted a novel type of materials, referred to as spin-gapless semiconductors (SGSs). The SGSs can be categorized as new members of the zero-gapless material family. Following this study, many Heusler-based SGSs [8] were studied. Recently, Zhang et al. studied the LuCoCrGe EQH compound [9], revealing that it is a highly dispersive gapless HM under strain.

Figure 1 illustrates a HM, SGS, and magnetic semiconductor (MS). Figure 1a shows that one of the channels (spin-down (minority) channel) is metallic, while the other one (spin-up (majority) channel)

exhibits a semiconducting behavior. Figure 1b shows that there is a band gap for the spin-up channel between the two bands (conduction and valence bands). The behavior is different for the spin-down channel, exhibiting a zero-gap between these bands; this behavior corresponds to SGS. In Figure 1c, two semiconductor-type band gaps are observed in both channels, corresponding to MS properties.

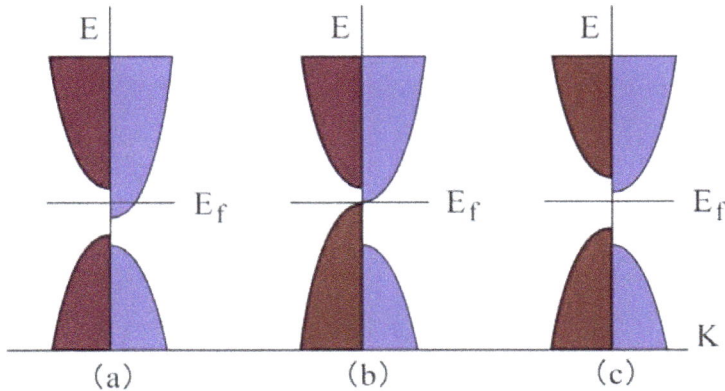

Figure 1. Schematics of the densities of states of a (**a**) half-metal (HM), (**b**) spin-gapless semiconductor (SGS), and (**c**) magnetic semiconductor (MS).

In this study, we employed first-principle calculations to investigate the crystal structure, electronic structure, and magnetic properties of a new rare-earth-containing EQH compound ScFeRhP. We demonstrate that this compound is HM at its equilibrium lattice constant. Rare physical transitions with an SGS feature can be observed at different strain magnitudes.

2. Calculation Method

We calculated the band structure and magnetic properties of ScFeRhP by the plane-wave pseudopotential method [10,11], using the Cambridge Serial Total-Energy Package (CASTEP) software. We studied the interaction between the valence electrons and nuclei by the method of ultra-soft pseudopotentials [12]. The generalized gradient approximation (GGA) was used to calculate the exchange and correlation between electrons [13] using the scheme of Perdew-Burke-Ernzerhof (PBE) [14]. In all calculations, a k-point mesh of $12 \times 12 \times 12$ and plane-wave basis-set cut-off of 450 eV were used. The above parameter settings ensure accuracy of the calculation results.

3. Results and Discussion

3.1. Structural Stability and Total Energy

In general, the EQH compounds have LiMgPdSn-type structures [15]. Three types of crystal structures and their distinct atomic positions of the ScFeRhP EQH compound are shown in Figure 2 and Table 1, respectively. Following the atom-occupation rule, Rh tends to occupy the site D (3/4, 3/4, 3/4), P and Sc tend to occupy the A (0, 0, 0) and C (1/2, 1/2, 1/2) sites, respectively, and Fe tends to occupy the B (1/4, 1/4, 1/4) site. Therefore, for the ScFeRhP compound, type I (see Figure 2) is the most probable configuration owing to its lowest energy.

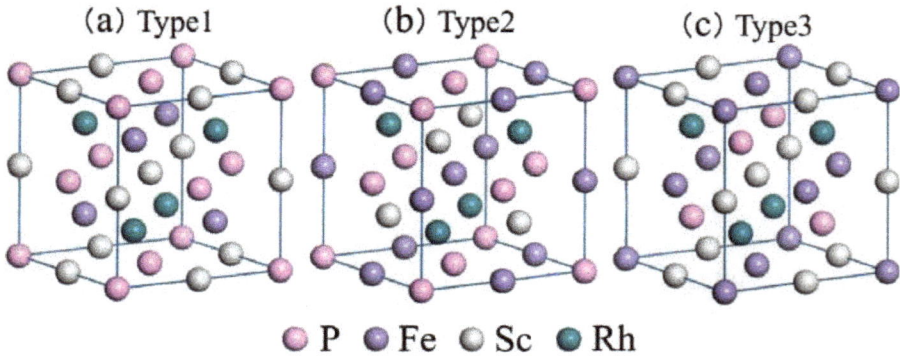

Figure 2. Three possible crystal structures (**a**) type 1, (**b**) type 2, (**c**) type 3 for ScFeRhP compound.

Table 1. Three crystal-structure atomic positions of the ScFeRhP equiatomic quaternary Heusler (EQH) compound.

Type	P	Fe	Sc	Rh
Type 1	A (0, 0, 0)	B (1/4, 1/4, 1/4)	C (1/2, 1/2, 1/2)	D (3/4, 3/4, 3/4)
Type 2	A (0, 0, 0)	C (1/2, 1/2, 1/2)	B (1/4, 1/4, 1/4)	D (3/4, 3/4, 3/4)
Type 3	B (1/4, 1/4, 1/4)	A (0, 0, 0)	C (1/2, 1/2, 1/2)	D (3/4, 3/4, 3/4)

For geometric optimization of the ScFeRhP EQH compound, the crystal cell energy is minimized as a function of the lattice constant. The three possible crystal structures are shown in Figure 2. Each of them has ferromagnetic (FM) and non-magnetic (NM) states. We calculated the energies by the CASTEP software. Figure 3a shows total energies in FM states for the type 1, 2, and 3 structures, respectively, and Figure 3b shows the energies of both FM state and NM state for type 1 structure. Among the considered cases the calculated result that the type 1 structure at the FM state exhibits the minimum energy. More details about the equilibrium lattice constants and the total energies of these three type structures can be seen in Table 2. This indicates that the most stable of these structures is the type 1 structure at the FM state. According to the calculations, the equilibrium lattice constant of the ScFeRhP EQH compound is 5.97 Å.

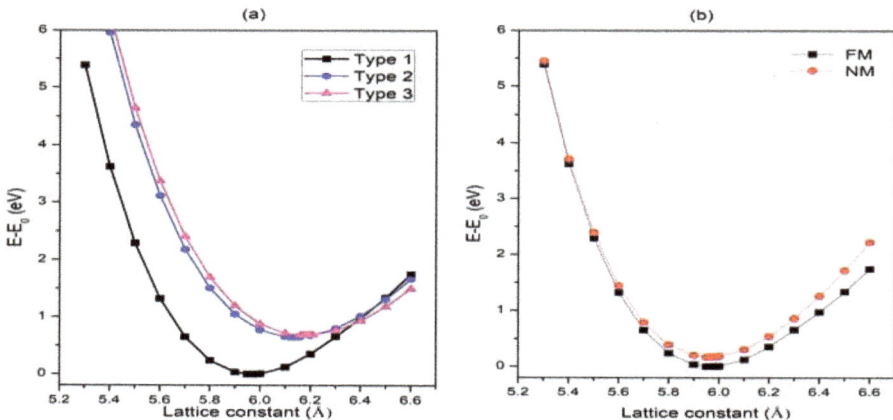

Figure 3. Total-energy-lattice-constant curves of the ScFeRhP compound in ferromagnetic (FM) states for the type (**a**) 1, 2, and 3 structures, and (**b**) FM and non-magnetic (NM) states for type 1.

Table 2. The equilibrium lattice constant and minimum total energy of each type for both FM and NM magnetic states.

States	Calculated Value	Type 1	Type 2	Type 3
FM	Equilibrium lattice constant	5.97 Å	6.14 Å	6.18 Å
	Total energy	−2931.40 eV	−2930.75 eV	−2930.72 eV
NM	Equilibrium lattice constant	5.96 Å	6.08 Å	6.10 Å
	Total energy	−2931.23 eV	−2929.80 eV	−2930.06 eV

3.2. Electronic Structure and Slater–Pauling Rule

The partial and total (M_t) element magnetic moments, and total number of valence electrons (Z_t) of this compound at the most stable configuration are shown in Table 3. The M_t of the ScFeRhP EQH compound is 1 μ_B, while its total number of valence electrons is 25. This EQH compound obeys the Slater-Pauling rule: $M_t = Z_t - 24$ [16].

Table 3. Partial and total magnetic moments (μ_B), calculated equilibrium lattice constant, Z_t, and Slater-Pauling (S-P) rule for the ScFeRhP compound.

Compound	Total	P	Fe	Sc	Rh	a (Å)	Z_t	S-P Rule
ScFeRhP	1.00	0.04	0.98	−0.24	0.22	5.97	25	$M_t = Z_t - 24$

Figure 4 shows the calculated band structures for the ScFeRhP compound. The band structure reveals the half-metallicity of this compound near the Fermi level. The spin-up channel exhibits a metallic character, while the spin-down channel exhibits a semiconducting character. Therefore, based on the obtained band structures and magnetism, we can conclude that the ScFeRhP EQH compound is a HM.

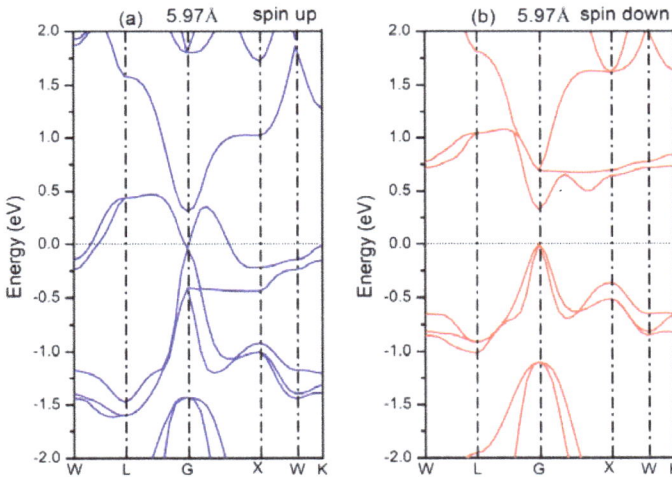

Figure 4. Band structures of the (**a**) spin-up and (**b**) spin-down channels of the ScFeRhP EQH compound at its equilibrium lattice constant.

We analyze the origin of the band gap at the spin-down channel in Figure 5. According to the study of Galanakis et al., P has completely occupied $1s$ and $3p$ states. We need to consider only the hybridization character of the $3d$ and $4d$ states of the ScFeRhP EQH compound, as shown in Figure 5. For this alloy, the d_4 and d_5 orbits of the Fe and Sc atoms couple forming antibonding e_u and bonding e_g states. The d_1, d_2, and d_3 orbits of the Fe and Sc atoms couple forming antibonding t_{1u} and bonding

t_{2g} states. The same states of the Rh atom hybridize with the above orbits, yielding 15 orbits (3 t_{2g}, 2 e_g, 2 e_u, 3 t_{1u}, 3 t_{2g}, and 2 e_g). There are 8 occupied orbits under the Fermi level (Figure 5). Combined with the 3p and 1s orbitals generated by P, there are a total of 12 orbits below the Fermi level; therefore, this compound follows the Slater-Pauling rule: $M_t = Z_t - 24$.

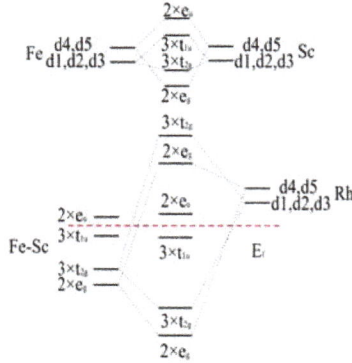

Figure 5. *d*-Orbit hybridization of the Rh (4*d*) and Fe/Sc (3*d*) transition elements in the ScFeRhP EQH compound.

In order to explain the origin of the band gap [17], we analyzed the partial densities of states (PDOS) and total density of states (TDOS) of ScFeRhP; the results are shown in Figure 6. The PDOS indicate that the TDOS at the Fermi level are mostly attributed to the Fe 3*d* and Rh 4*d* states. The PDOS of the P atoms and rare-earth element Sc are significantly lower near the Fermi level compared with those of the Fe and Rh atoms. Figure 6 reveals that the Fe element exhibits a stronger spin splitting at −1.4 eV in the majority channel and −0.8 eV in the other channel for the bonding state. For the Rh element, the bonding state was observed mostly in the range of −3.5 eV to −4 eV in the majority channel and in the range of −3 eV to −3.5 eV in the other channel.

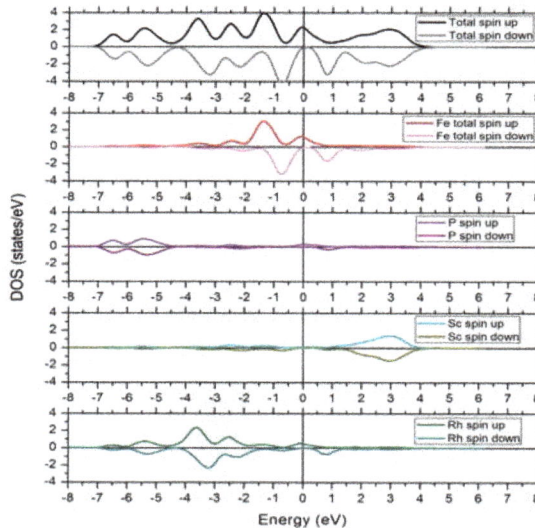

Figure 6. Total density of states (TDOS) and partial densities of states (PDOSs) for ScFeRhP.

3.3. Magnetic Properties

The magnetic behavior of the ScFeRhP alloy at strained lattice constants is discussed in detail in this section. The M_t of the ScFeRhP EQH compound is 1 μ_B at its equilibrium lattice constant, and it remains almost unchanged when the lattice constant changed in a large range. The main contribution to the magnetic moment originates from the Fe atoms, as shown in Table 3. The M_t and partial magnetic moments at strained lattice constants of the ScFeRhP compound are presented in Figure 7. According to the actual conditions, we focus on compressive and expanded lattice constants in the range of 5.30 Å to 6.10 Å. As shown above, the total magnetic moment of the ScFeRhP EQH compound is 1 μ_B at its equilibrium lattice constant. The magnetic moment decreases with the increase of the lattice constant for both P and Sc atoms; the magnetic moment of the Fe atom continuously increases, while that of the Rh atom is almost constant.

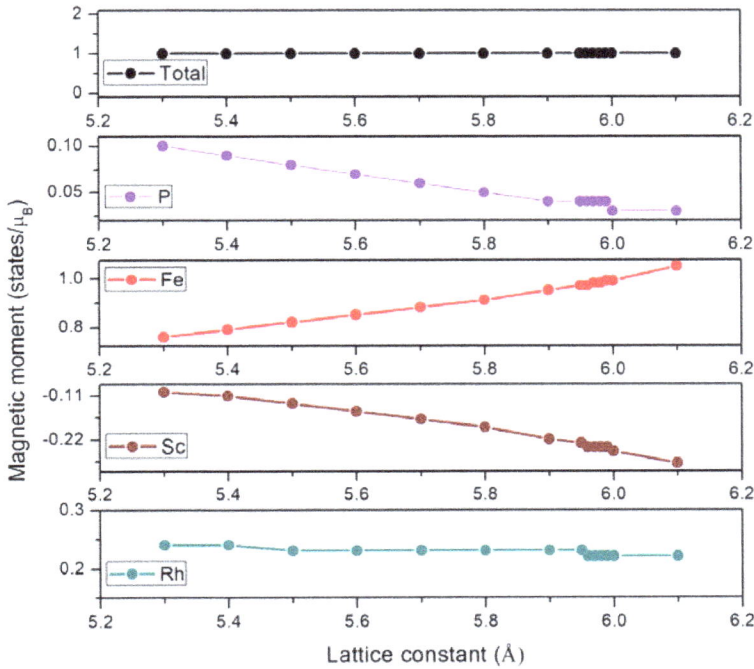

Figure 7. Partial and total magnetic moments as a function of the lattice constant of the ScFeRhP EQH compound.

3.4. Transitions of the Physical State under Uniform Strain

In this section, we discuss the change of the physical state of the compound under uniform strain. A novel distinct transition can be observed in the obtained band structure of the ScFeRhP compound at its strained lattice constants. The conversion of non-MS-(NMS)-to-MS-to-SGS-to-HM-to-M for this compound is shown in Figure 8.

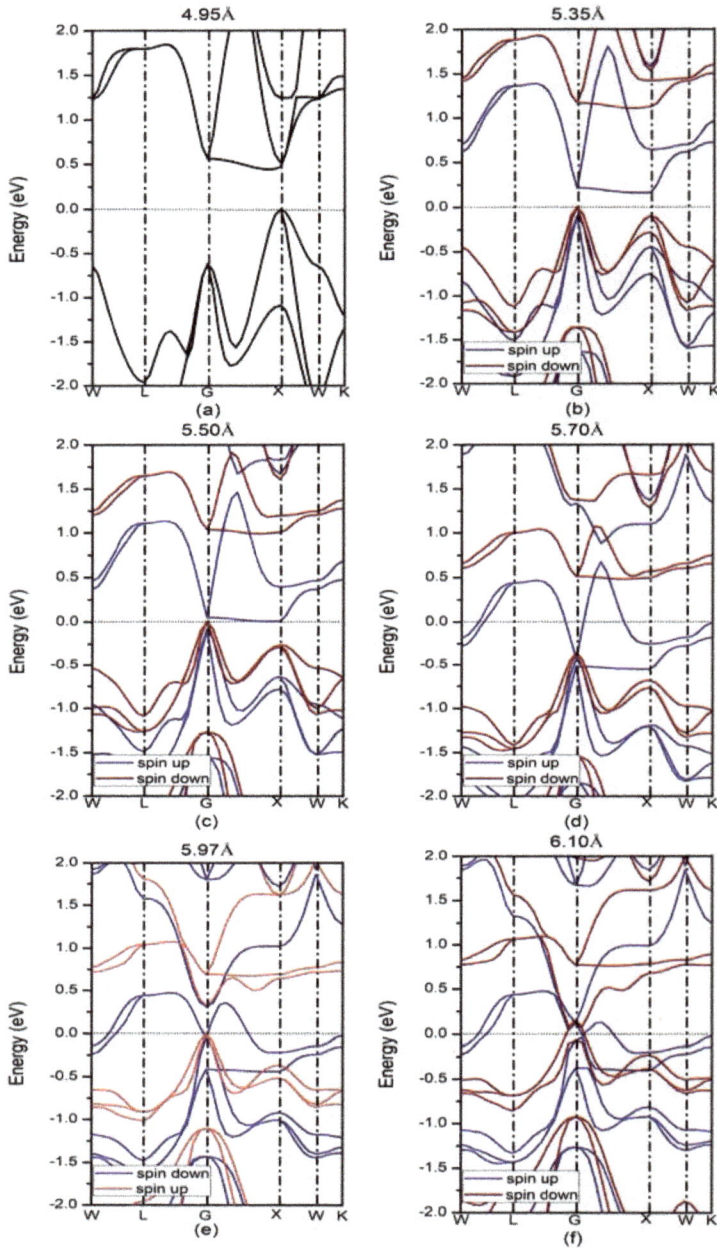

Figure 8. (**a–f**) are band structures under strained lattice constants of 4.95 Å, 5.35 Å, 5.50 Å, 5.70 Å, 5.97 Å, and 6.10 Å, respectively, for ScFeRhP compound.

The detailed band structures are plotted in Figure 8. With the increase of the lattice constant, the valence band moves down in the minority channel at the *X*-point, and moves up in the majority channel at the *G*-point, whereas in the conduction band, the opposite behavior is observed. If the lattice constant is smaller than 4.95 Å, the compound is NMS. If the parameter is in the range of 4.95 Å to 5.495 Å, the compound is MS. If the parameter is in the range of 5.495 Å to 5.505 Å, a zero-gap between the spin-up channel in the valence band and spin-down channel in the conduction band appears; the compound is SGS. If the parameter is in the range of 5.505 Å to 6.05 Å, the compound is HM, while for lattice constants larger than 6.05 Å, the compound is metal. These results are illustrated in Figure 9.

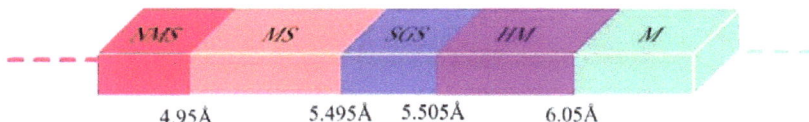

Figure 9. Physical transitions under uniformly strained lattice constants.

4. Conclusions

Using first-principle calculations, the crystal structure, band structure, magnetic properties, and origin of the band gap of the ScFeRhP compound were studied. The compound is HM at its equilibrium lattice constant. Upon introduction of strain, the compound exhibited transitions of its physic state (NMS → MS → SGS → HM → M), which implies that the magnetic properties and electronic structure could be widely changed by external tension or compression. A SGS feature appeared by tuning the lattice constant of the ScFeRhP compound. This study indicates that the ScFeRhP EQH compound can be used in spintronic applications.

Author Contributions: Y.G. conceived and designed the studies; Z.C. performed the calculations; H.R. and H.X. analyzed the data; Z.C. wrote the paper.

Funding: This research received no external funding.

Conflicts of Interest: The authors declare there is no conflicts of interest regarding the publication of this paper.

References

1. Groot, R.A.D.; Mueller, F.M.; Engen, P.G.V.; Buschow, K.H.J. New class of materials: Half-metallic ferromagnets. *Phys. Rev. Lett.* **1983**, *50*, 2024–2027. [CrossRef]
2. Wang, X.T.; Dai, X.F.; Wang, L.Y.; Liu, X.F.; Wang, W.H.; Wu, G.H.; Tang, C.C.; Liu, G.D. Electronic structures and magnetism of Rh_3Z (Z = Al, Ga, In, Si, Ge, Sn, Pb, Sb) with DO_3 structures. *J. Magn. Magn. Mater.* **2015**, *378*, 16–23. [CrossRef]
3. Wang, X.; Cheng, Z.; Liu, G. Largest magnetic moments in the half-Heusler alloys XCrZ (X = Li, K, Rb, Cs; Z = S, Se, Te): A first-principles study. *Materials* **2017**, *10*, 1078. [CrossRef] [PubMed]
4. Wang, X.; Cheng, Z.; Wang, W. $L2_1$ and XA ordering competition in hafnium-based full-Heusler alloys Hf_2VZ (Z = Al, Ga, In, Tl, Si, Ge, Sn, Pb). *Materials* **2017**, *10*, 1200. [CrossRef] [PubMed]
5. Bainsla, L.; Suresh, K.G. Equiatomic quaternary Heusler alloys: A material perspective for spintronic applications. *Appl. Phys. Rev.* **2016**, *3*, 031101. [CrossRef]
6. Han, Y.; Wu, Y.; Li, T.; Khenata, R.; Yang, T.; Wang, X. Electronic, magnetic, half-metallic, and mechanical properties of a new equiatomic quaternary Heusler compound YRhTiGe: A first-principles study. *Materials* **2018**, *11*, 797. [CrossRef] [PubMed]
7. Wang, X.L. Proposal for a new class of materials: Spin gapless semiconductors. *Phys. Rev. Lett.* **2008**, *100*, 156404. [CrossRef] [PubMed]
8. Wang, X.; Cheng, Z.; Liu, G.; Dai, X.; Khenata, R.; Wang, L.; Bouhemadou, A. Rare earth-based quaternary Heusler compounds MCoVZ (M = Lu, Y; Z = Si, Ge) with tunable band characteristics for potential spintronic applications. *IUCrJ* **2017**, *4*, 758–768. [CrossRef] [PubMed]
9. Zhang, L.; Wang, X.; Cheng, Z. Electronic, magnetic, mechanical, half-metallic and highly dispersive zero-gap half-metallic properties of rare-earth-element-based quaternary Heusler compounds. *J. Alloys Compd.* **2017**, *718*, 63–74. [CrossRef]

10. Payne, M.C.; Teter, M.P.; Allan, D.C.; Arias, T.A.; Joannopoulos, J.D. Iterative minimization techniques for ab initio total-energy calculations: Molecular dynamics and conjugate gradients. *Rev. Mod. Phys.* **1992**, *64*, 1045–1097. [CrossRef]

11. Segall, M.D.; Lindan, P.J.; Probert, M.A.; Pickard, C.J.; Hasnip, P.J.; Clark, S.J.; Payne, M.C. First-principles simulation: Ideas, illustrations and the CASTEP code. *J. Phys.* **2002**, *14*, 2717. [CrossRef]

12. Vanderbilt, D. Soft self-consistent pseudopotentials in a generalized eigenvalue formalism. *Phys. Rev. B* **1990**, *41*, 7892. [CrossRef]

13. Perdew, J.P.; Burke, K.; Ernzerhof, M. Generalized gradient approximation made simple. *Phys. Rev. Lett.* **1996**, *77*, 3865–3868. [CrossRef] [PubMed]

14. Maximoff, S.N.; Ernzerhof, M.; Scuseria, G.E. Current-dependent extension of the Perdew-Burke-Ernzerhof exchange-correlation functional. *J. Chem. Phys.* **2004**, *120*, 2105–2109. [CrossRef] [PubMed]

15. Ozdogan, K.; Sasıoglu, E.; Galanakis, I. Slater-Pauling behavior in LiMgPdSn-type multifunctional quaternary Heusler materials: Half-metallicity, spin-gapless and magnetic semiconductors. *J. Appl. Phys.* **2013**, *113*, 193903. [CrossRef]

16. Galanakis, I.; Dederichs, P.H.; Papanikolaou, N. Slater-Pauling behavior and origin of the half-metallicity of the full-Heusler alloys. *Phys. Rev. B* **2002**, *66*, 174429. [CrossRef]

17. Galanakis, I.; Mavropoulos, P.; Dederichs, P.H. Electronic structure and Slater-Pauling behaviour in half-metallic Heusler alloys calculated from first principles. *J. Phys. D Appl. Phys.* **2006**, *39*, 765. [CrossRef]

applied
sciences

MDPI

Article

Interface Characterization of Current-Perpendicular-to-Plane Spin Valves Based on Spin Gapless Semiconductor Mn₂CoAl

Ming-Sheng Wei, Zhou Cui, Xin Ruan, Qi-Wen Zhou, Xiao-Yi Fu, Zhen-Yan Liu, Qian-Ya Ma and Yu Feng *

School of Physics and Electronic Engineering, Jiangsu Normal University, Xuzhou 221116, China; weims@jsnu.edu.cn (M.-S.W.); cuizhoujsnu@163.com (Z.C.); ruanxinjsnu@163.com (X.R.); zhouqiwenjsnu@163.com (Q.-W.Z.); fuxiaoyijsnu@163.com (X.-Y.F.); liuzhenyanjsnu@163.com (Z.-Y.L.); maqianyajsnu@163.com (Q.-Y.M.)
* Correspondence: fengyu@jsnu.edu.cn; Tel.: +86-516-835-00485

Received: 19 July 2018; Accepted: 9 August 2018; Published: 10 August 2018

Featured Application: Current-perpendicular-to-plane spin valve based on spin gapless semiconductor Mn₂CoAl with MnCo termination possesses high value of magnetoresistance of 2886% and has a better application in a spintronics device.

Abstract: Employing the first-principles calculations within density functional theory (DFT) combined with the nonequilibrium Green's function, we investigated the interfacial electronic, magnetic, and spin transport properties of Mn₂CoAl/Ag/Mn₂CoAl current-perpendicular-to-plane spin valves (CPP-SV). Due to the interface rehybridization, the magnetic moment of the interface atom gets enhanced. Further analysis on electronic structures reveals that owing to the interface states, the interface spin polarization is decreased. The largest interface spin polarization (ISP) of 78% belongs to the MnCo^T-terminated interface, and the ISP of the MnMn^{T1}-terminated interface is also as high as 45%. The transmission curves of Mn₂CoAl/Ag/Mn₂CoAl reveal that the transmission coefficient at the Fermi level in the majority spin channel is much higher than that in the minority spin channel. Furthermore, the calculated magnetoresistance (MR) ratio of the MnCo^T-terminated interface reaches up to 2886%, while that of the MnMn^{T1}-terminated interface is only 330%. Therefore, Mn₂CoAl/Ag/Mn₂CoAl CPP-SV with an MnCo-terminated interface structure has a better application in a spintronics device.

Keywords: Heusler alloy; spin gapless semiconductor; electronic structure; spin transport

1. Introduction

The current-perpendicular-to-plane spin valve (CPP-SV) is regarded as one of the most significant spintronics devices, and has great potential for spin transfer torque devices in spin random access memory and ultra-high-speed reading in magnetic read heads of hard disk drivers [1–4]. The generation and detection of highly polarized spin current is one of the foremost challenges of a CPP-SV. Half metallic Heusler alloys (HMHA) usually have a high Curie temperature (T_C) and their lattice constants are close to Ag and Cu. In addition, one spin band (majority spin) of HMHA exhibits metallic behavior, while the other spin band (minority spin) displays semiconductor behavior, and such a band structure could generate a high spin polarized current. Therefore, HMHA has been explored as an electrode of CPP-SV to achieve a high magnetoresistance (MR) ratio [5–11]. Takayashi et al. reported a large MR ratio of 74.8% at room temperature (RT) in CPP-SV using Co₂MnGa₀.₂₅Ge₀.₇₅ [12]. Sakuraba et al. demonstrated an MR ratio of 58% at RT and 184% at 30 K in Co₂Fe₀.₄Mn₀.₆Si/Ag

CPP-SV [13]. Moreover, a number of CPP-SVs employing a Heusler alloy as the spin injector, such as Co_2MnSi/Ag, $Co_2FeAl_{0.5}Si_{0.5}/Ag$, and $Co_2FeGa_{0.5}Ge_{0.5}/Ag$, also obtained high MR ratios [14,15]. More recently, the Hg_2CuTi type Heusler alloy Mn_2CoAl has been demonstrated to be one of the spin gapless semiconductors (SGS) from theoretical calculation combined with experimental work [16–26]. The minority spin band of Mn_2CoAl has an energy gap around the Fermi level, while the top of the valence band and the bottom of the conduction band touch at the Fermi level (see Figure 1). The special band structure allows the holes and electrons to be spin polarized simultaneously. Mn_2CoAl has attracted much attention and it is considered to possess great potential for spintronics. Since the MR ratio of spin valves based on a Heusler alloy severely depends on its interface properties, and in order to explore outstanding candidates for CPP-SV, we study the interface magnetism, electronic structure, and spin transport properties of $Mn_2CoAl/Ag/Mn_2CoAl$ CPP-SV.

There are two ideal terminations along the (0 0 1) direction in a Heusler alloy Mn_2CoAl: MnCo and MnAl. Besides, there are two ways to connect a Heusler alloy with an Ag spacer, one of which is where atoms from the Heusler alloy sit on top of Ag atoms, which is noted as $MnCo^T$ and $MnAl^T$. The other way is where atoms from the Heusler alloy locate in the bridge site between Ag atoms, which is noted as $MnCo^B$ and $MnAl^B$. In addition, according to previous works, a pure Mn-modified artificial termination could improve the spin polarization [27,28], so MnMn terminations are also considered. When interface Co atoms in $MnCo^T$ and $MnCo^B$ are replaced by Mn atoms, $MnMn^{T1}$ and $MnMn^{B1}$ are built, respectively. Similarly, when interface Al atoms in $MnAl^T$ and $MnAl^B$ are substituted with Mn atoms, $MnMn^{T2}$ and $MnMn^{B2}$ are built, respectively. Various interface structures of $Mn_2CoAl/Ag/Mn_2CoAl$ CPP-SV are exhibited in Figure 2.

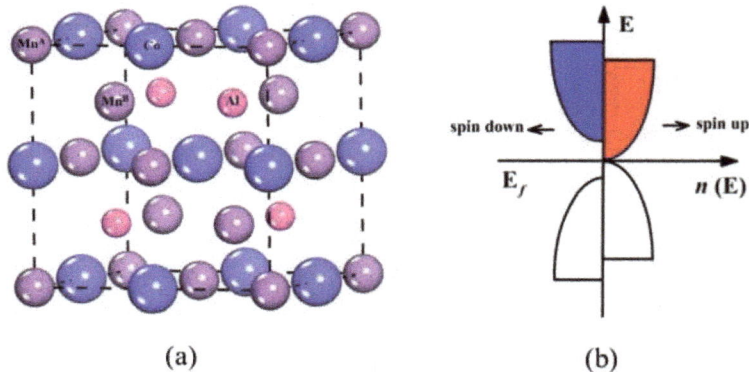

Figure 1. (**a**) Schematic representation of Mn_2CoAl bulk structure. The schematic density of states of n (E) as a function of energy E is shown for (**b**) a spin gapless semiconductor. The occupied states are indicated by filled areas. Arrows indicate the majority and minority states.

2. Calculation Methods

The structure optimization, electronic, and magnetic properties are calculated by utilizing the Vienna ab initio Simulation Package (VASP) based on density functional theory (DFT) within generalized gradient approximation (GGA) [29]. We make use of projector-augmental wave (PAW) [30] pseudopotential to deal with the electron-iron interaction, and the Mn ($3d^54s^2$), Co ($3d^74s^2$), Al ($3s^23p^1$), and Ag ($4d^{10}5s^1$) are selected as valence-electron configurations. The mesh of special k-points in the Brillouin zone is set to $7 \times 7 \times 1$ and the SCF convergence criterion of 10^{-6} eV is applied. The cutoff energy is taken to be 500 eV. The optimized Ag and Mn_2CoAl bulks are cleaved along the Miller indices (0 0 1) crystal direction, and we chose 13 Mn_2CoAl layers as the electrode and nine Ag layers as the spacer layer to compose the $Mn_2CoAl/Ag/Mn_2CoAl$ heterojunction. The in-plane lattice parameter is fixed to 4.1 Å, which corresponds to $1/\sqrt{2}$ of the experimental lattice constant of Mn_2CoAl bulk (5.798 Å [20]), and such a value is very close to the lattice constant of Ag. Therefore, the lattice mismatch between the Mn_2CoAl electrode and Ag is very small. The spin-dependent transport properties calculations are based on a state-of-the-art technique where DFT is combined with the Keldysh nonequilibrium Green's function (NEGF) theory, as implemented in the Nanodcal package [31,32].

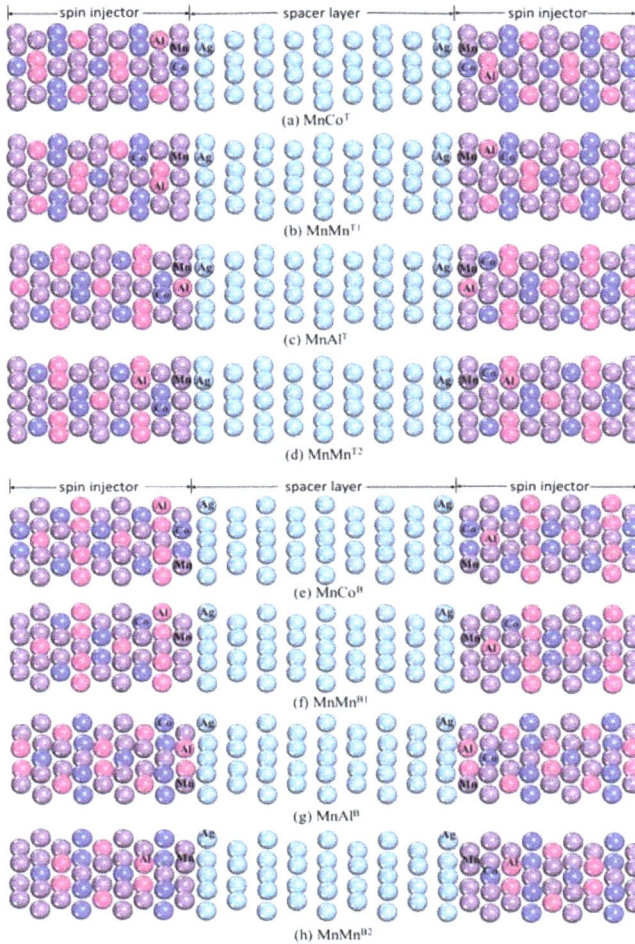

Figure 2. Various atomic terminations of the $Mn_2CoAl/Ag/Mn_2CoAl$ (001) interface.

3. Result and Discussion

The displacements of interface atoms in various $Mn_2CoAl/Ag/Mn_2CoAl$ structures are measured and exhibited in Figure 3. L1 and L2 indicate the interface and subinterface, respectively. It can be seen that the interfacial Mn^A atom has an obvious outward extension in $MnCo^B$, $MnCo^T$, $MnMn^{T1}$, and $MnMn^{B1}$ structures, revealing that the interface Mn atom prefers to bond with the interface atom from the spacer layer, and such behavior is also reported in previous theoretical and experimental studies [33,34]. As for the interfacial Mn^B atom, it slightly stretches outward in $MnMn^{T2}$ and $MnMn^{B2}$, and keeps its ideal position in MnAl^T, while it even shrinks inward in MnAl^B, showing that its bonding ability is weaker than interfacial Mn^A. The interface Co atom has a slight outward movement, and when the MnMn^{T1} and MnMn^{B1}-terminated interface are respectively formed by substituting interface Co with the Mn atom, such an outward action is enhanced. Besides, for various different structures, the outward movement of the subinterface atom is very small, and the position of the next subinterface atom is nearly unchanged, so their behaviors are extremely similar to the bulk (not shown in Figure 3).

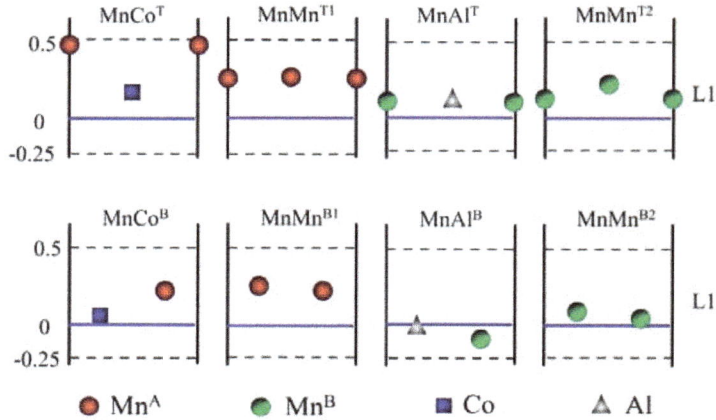

Figure 3. The relaxed atomic positions in various terminations. L1 and L2 indicate the interface and subinterface, respectively.

The interface free energy (γ) versus atomic chemical potentials is calculated in order to determine the interface stability. As was mentioned in Ref. [34], the γ can be expressed as

$$\gamma = \frac{1}{2A}\left[G - \sum_i \mu_i N_i\right] \tag{1}$$

where A is the interface area of a supercell, G stands for the Gibbs free energy of the slab; and μ_i and N_i are the chemical potential and the number of the ith atom, respectively. The relationship among the interdependent chemical potentials can be expressed as $2\mu_{Mn} + \mu_{Co} + \mu_{Al} = G_{bulk}$. Besides, the highest values of μ_{Co} and μ_{Mn} are $\mu_{Co} = G_{Co}$ and $\mu_{Mn} = G_{Mn}$, where G_{Co} and G_{Mn} are Gibbs free energies of the Co and Mn bulk, respectively. Besides, due to the reason that 2Mn + CoAl = Mn$_2$CoAl and Co + Mn$_2$Al = Mn$_2$CoAl, the lowest values of μ_{Co} and μ_{Mn} are

$$\mu_{Mn} = \frac{1}{2}\left[G_{bulk} - G_{CoAl}\right], \quad \mu_{Co} = \frac{1}{2}\left[G_{bulk} - G_{Mn_2Al}\right] \tag{2}$$

In Figure 4, the MnMn^{T1} and MnMn^{B1} always stay in the thermodynamic accessible region (TAR) and occupy the maximum area. As μ_{Mn} and μ_{Co} increase, MnMn^{T2}, MnMn^{B2}, MnCo^T, and MnCo^B are included in the TAR, revealing that these interfaces could be stable under the condition of Mn rich and Co rich. In addition, MnMn^{B2} enters into the TAR under the condition of $\mu_{Co} = -0.702$ Ryd and $\mu_{Mn} = -0.523$ Ryd, while MnMn^{T2} appears in the TAR when $\mu_{Co} = -0.702$ Ryd and $\mu_{Mn} = -0.694$ Ryd, indicating that the formation of a stable MnMn^{T2} needs a richer Mn condition than MnMn^{B2}. In the case of MnCo^B, it enters into the TAR when μ_{Co} and μ_{Mn} respectively increase to -0.672 Ryd and -0.477 Ryd, while MnCo^T appears in the TAR when $\mu_{Co} = -0.68$ Ryd and $\mu_{Mn} = -0.547$ Ryd. Therefore, it could be deduced that the MnCo terminated interface can be more easily obtained in a bridge-type

Appl. Sci. **2018**, *8*, 1348

structure than in a top-type structure. However, as for MnAl terminated interfaces, they are excluded from the TAR in both top-type and bridge-type structures.

The atom-resolved spin magnetic moments (AMMs) in the first three layers are plotted in Figure 5. In the Mn_2CoAl bulk, each atom locates at a tetrahedral symmetry position, and each Co and Mn^A atom has four Al and four Mn^B as the nearest neighbors. While in the $Mn_2CoAl/Ag/Mn_2CoAl$ heterostructures, an Ag atom takes the place of half of the nearest neighbors of the interface atom, reducing atomic coordination numbers at the interface. As a result, it breaks the periodic crystal field at the interface and the localization of the d-electron atom at the interface is enhanced. Therefore, AMMs of interface atoms are strengthened in various structures, and are larger than their corresponding bulk values. Besides, the relative distance between magnetic atoms could also affect the value of AMM. The maximum outward extension belongs to the interface Mn atom in the $MnCo^T$ structure, and it results in reduced hybridization and strengthened localization. Therefore, its AMM rises up to the highest value of 3.73 μ_B. The interface Mn atom in $MnMn^{B1}$ has a lower AMM than in $MnCo^B$, although they have a similar outward movement. This could be ascribed to the fact that the interface of $MnMn^{B1}$ is completely occupied by Mn atoms, and the interaction between Mn atoms is stronger than that between Co and Mn atoms. Such an explanation is also applied to the interface Mn atom in $MnAl^T$ and $MnMn^{T2}$, as well as $MnAl^B$ and $MnMn^{B2}$.

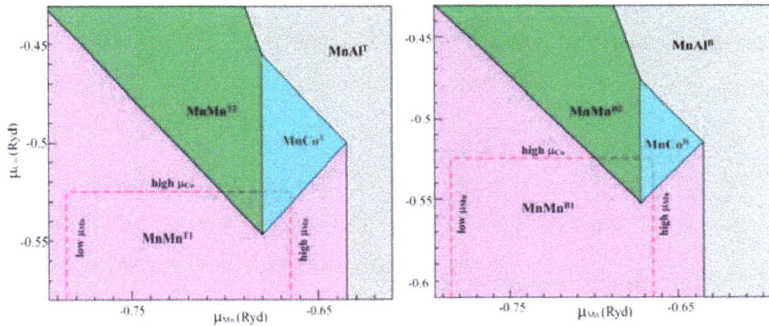

Figure 4. The calculated phase diagram of the $Mn_2CoAl/Ag/Mn_2CoAl$ (001) interface. The dashed lines denote the thermodynamically accessible region.

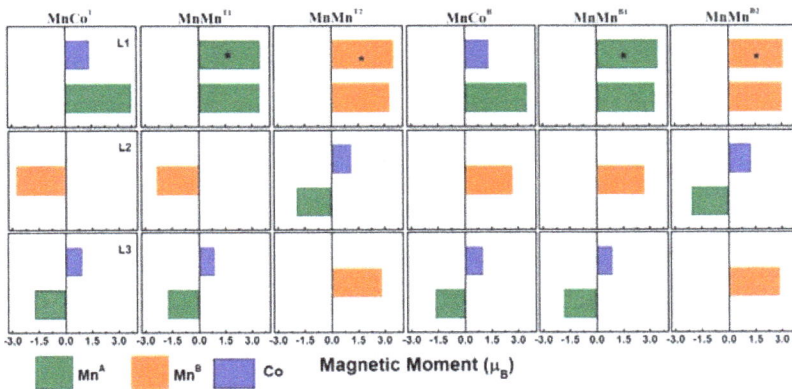

Figure 5. Atom-resolved spin magnetic moment of magnetic atoms of various terminations. L1, L2, and L3 indicate the interface, subinterface, and next subinterface, respectively. The region with '*' stands for the replaced atom.

Owing to the fact that subinterface atoms have small displacement, and the atom at the next subinterface maintains its position, AMMs of the subinterface and the next subinterface atoms could be compared to corresponding values in the bulk, revealing that atoms at deeper layers receive a minor interface effect. AMM of the Al atom presents negative values due to the fact that the *sp*-atom has a function of a bridge between local *d*-electron atoms according to the RKKY exchange model.

Spin transportation of a spintronics device severely depends on interface spin polarization (ISP), which can be written as $(D_\uparrow - D_\downarrow)/(D_\uparrow + D_\downarrow)$, where the $D_\uparrow$ and $D_\downarrow$ respectively represent the majority spin density of states and minority spin density of states. First, we calculate the total density of states (DOS). It can be seen from Figure 6 that interface half-metallicity is not detected in various structures. The MnCo^T interface displays the highest ISP of 78%, and the ISP of MnMn^{T1} also exhibits a value as high as 45%; while the ISP of MnMn^{T2}, MnCo^T, and MnMn^{B1} are only 26%, 33%, and 32%, respectively. The ISP of MnMn^{B2} even descends to a poor value of 10%. Hence, it reveals that with different Co$_2$MnSi/GaAs and CoFeMnSi/GaAs heterojunctions [27,35], the interface containing the pure Mn atom in Mn$_2$CoAl/Ag spin valves could not improve the ISP.

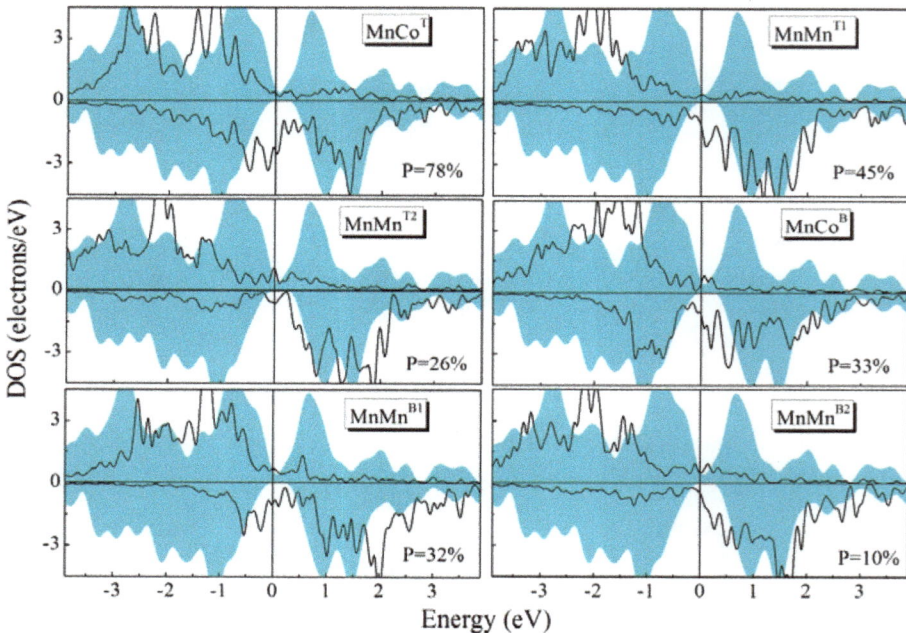

Figure 6. The total density of states (DOS) of various structures. The shadow region indicates the DOS of the Mn$_2$CoAl bulk. P stands for spin polarization.

Due to the reason that MnCo^T and MnMn^{T1} structures possess a high ISP, the project density of states (PDOS) in the interface, subinterface, and the next subinterface of these two terminated structures are further investigated and displayed in Figure 7. As for the interface magnetic atom, its majority spin band shifts to the lower energy region, while the minority spin band moves towards the higher energy region. Therefore, the exchange splitting is increased and magnetic moments are increased owing to the truncated periodic crystal field at the interface. The Mn^A atom in Mn$_2$CoAl bulk displays reverse splitting, which leads to its negative value of magnetic moment, while it converts to normal splitting at the interface, and its magnetic moment obtains a positive value. Quite the opposite, the normal splitting of the Mn^B atom in the bulk shifts to reverse splitting at the subinterface of MnCo^T

Appl. Sci. **2018**, *8*, 1348

and MnMn^{T1} structures. It indicates that the antiferromagnetic interaction of Mn^A in bulk turns to a ferromagnetic interaction when it at the interface, while the ferromagnetic coupling of bulk Mn^B converts to antiferromagnetic coupling when at the subinterface. Besides, some polarized peaks which have a *d*-orbital characteristic appear in the minority spin energy gap at the interface, and the energy of these peaks decreases in the second layer and finally vanishes in the third layer. Therefore, these polarized peaks are considered to be interface states, and they are responsible for the destruction of interface half-metallicity. In addition, it can be seen that the PDOSs of atoms at the third layer are very similar to their features in bulk, revealing that the electronic structures of the deeper layer receive a minor influence of the interface effect.

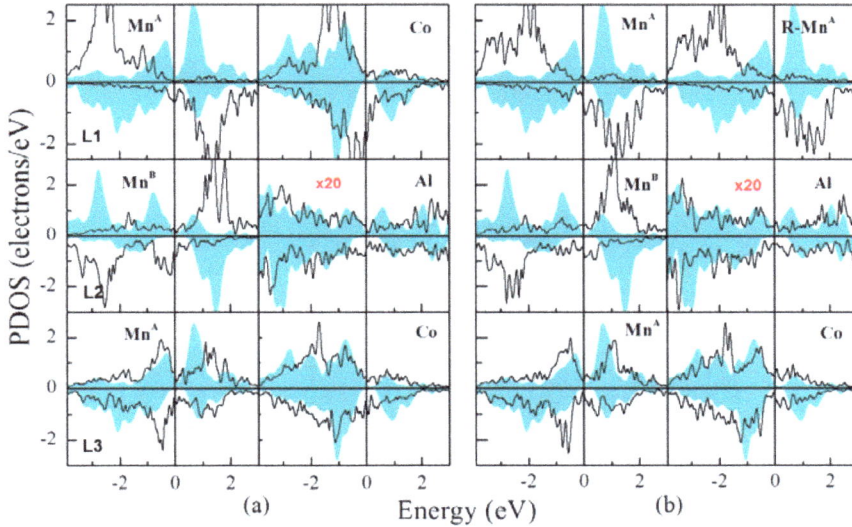

Figure 7. The partial density of states (PDOS) of (**a**) MnCo^T- and (**b**) MnMn^{T1}-terminated structures. The shadow region indicates atomic PDOS of the bulk. L1, L2, and L3 indicate the interface, subinterface, and next subinterface, respectively.

Among various investigated structures, due to the reason that the ISP of the MnCo^T-terminated and MnMn^{T1}-terminated interface reaches up to 78% and 45%, respectively; the transport properties of the Mn$_2$CoAl/Ag/Mn$_2$CoAl spin valve with an MnCo^T-terminated and MnMn^{T1}-terminated interface have been calculated. The energy- and spin-resolved transmission coefficient $T^\sigma(E)$ is defined as

$$T^\sigma(E) = \frac{1}{n^2} \int d^2k_{//} T^\sigma(\vec{k}_{//}, E) \tag{3}$$

where n^2 represents the number of sampling points in the two-dimensional Brillouin zone (2-D BZ), and σ is the spin direction. The transmission coefficients of the Mn$_2$CoAl/Ag/Mn$_2$CoAl junction with an MnCo^T-terminated and MnMn^{T1}-terminated interface at equilibrium are calculated and presented in Figure 8, and the Fermi level has an energy of zero. When magnetic moments of two Mn$_2$CoAl electrode layers are in a parallel configuration (PC), the transmission coefficient curve of the majority spin channel is totally different to that of the minority spin channel. For the MnCo^T-terminated structure, the value of the transmission coefficient of the majority spin channel at Fermi level $T_{PC}^{maj}(E_f)$ is 0.00663, and that of the minority spin channel at Fermi level $T_{PC}^{min}(E_f)$ is almost zero. For the MnMn^{T1}-terminated structure, $T_{PC}^{maj}(E_f)$ is 0.0038 and $T_{PC}^{min}(E_f)$ is also close to zero. When the magnetic moments of two Mn$_2$CoAl electrode layers are in an anti-parallel configuration (APC), due to the reason that our $Mn_2CoAl/Ag/Mn_2CoAl$ spin valve model is geometrically symmetric with respect to the middle plane of the center scattering region, the transmission coefficient curve of the spin up channel is exactly the same as that of the spin down channel. For the $MnCo^T$-terminated structure, the value of the transmission coefficient of the majority spin channel and minority spin channel at the Fermi level is $T_{APC}^{maj}(E_f) = T_{APC}^{min}(E_f) = 0.111 \times 10^{-3}$. For the $MnMn^{T1}$-terminated structure, $T_{APC}^{maj}(E_f) = T_{APC}^{min}(E_f) = 0.442 \times 10^{-3}$.

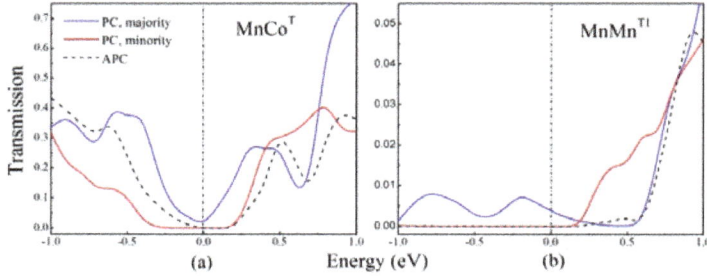

Figure 8. Transmission coefficient versus electron energy in PC (parallel configuration) and APC (anti-parallel configuration) of the $Mn_2CoAl/Ag/Mn_2CoAl$ CPP-SV at equilibrium; (**a**) The structure with $MnCo^T$-termination; (**b**) The structure with $MnMn^{T1}$-termination.

The magnetoresistance (MR) ratio of $Mn_2CoAl/Ag/Mn_2CoAl$ CPP-SV with an $MnCo^T$-terminated and $MnMn^{T1}$-terminated interface is also calculated by:

$$MR = \left| \frac{G_{pc} - G_{APC}}{\min(G_{pc}, G_{APC})} \right| \times 100\% \tag{4}$$

where G_{pc} and G_{Apc} are conductance at the Fermi level in PC and APC, respectively, where $G_{PC} = T_{PC}^{maj} + T_{PC}^{min}$ and $G_{APC} = T_{APC}^{maj} + T_{APC}^{min}$. According to the calculated values in Figure 8, the MR ratio of the $MnCo^T$-terminated structure is as high as 2886%, while that of the $MnMn^{T1}$-terminated structure is only about 330%. Our calculated results indicate that the $Mn_2CoAl/Ag/Mn_2CoAl$ spin valve with the $MnCo^T$-terminated interface is a high-performance spintronics device. Furthermore, the 2-D BZ transmission coefficients at the Fermi level as a function of k_x and k_y, which are perpendicular to the transport direction (z axis), have been calculated at equilibrium. The contour plots of the $\overrightarrow{k}_{//}$ dependence of the spin up and spin down transmission coefficients in PC and APC have been mapped in Figure 9. It can be seen that the density of dark-red spots in Figure 9a is much higher than that in Figure 9b, indicating that the majority spin electrons in PC have larger transmission probabilities than minority spin electrons, and it is consistent with the calculated transmission curves exhibited in Figure 8. In addition, the contour plots of the majority spin transmission coefficient in APC are exactly the same as those of the minority spin transmission coefficient, and rare *hot spots* exist in Figure 9c,d, indicating that the transmissions of both majority spin and minority spin channels in APC have been suppressed.

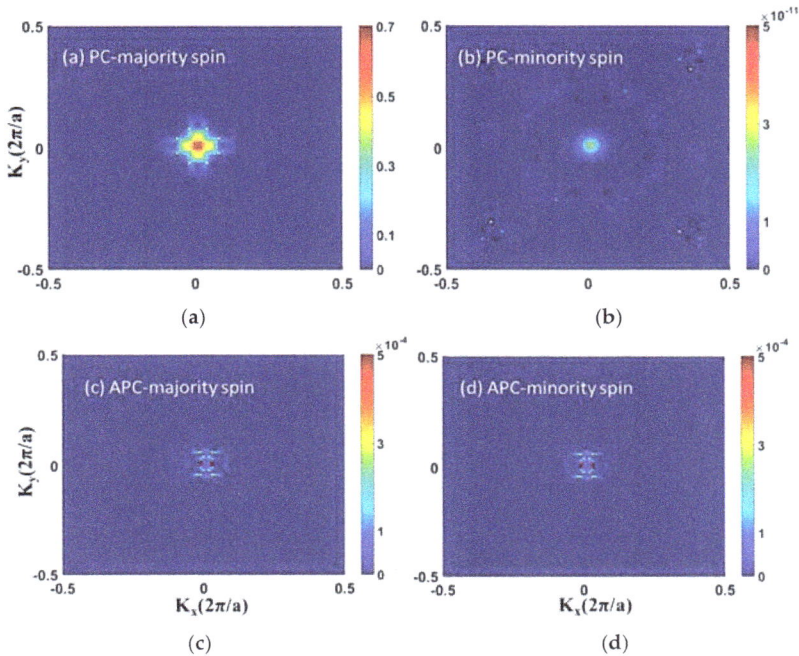

Figure 9. The $k_{//}$-resolved transmission coefficients of the MnCo^T-terminated structure at the Fermi level $E = E_f$. (**a**) Majoirty spin in PC; (**b**) minority spin in PC; (**c**) majority spin in APC; (**d**) minority spin in APC.

4. Conclusions

Combining the first-principles calculations within density functional theory (DFT) with nonequilibrium Green's function, we investigated the interface magnetism and electronic structures of Mn$_2$CoAl/Ag/Mn$_2$CoAl CPP-SV. Our calculations reveal that due to the reason that crystal symmetry at the interface is broken, the rehybridization leading to the AMMs of atoms at the interface becomes enhanced. Moreover, analyses of electronic structures reveal that owing to the interface states, the interface half-metallicity is destroyed. The MnCo^T-terminated interface preserves the highest ISP of 78%, and the MnMn^{T1}-terminated interface also has a high ISP of 45%. The transmission curves reveal that the transmission coefficient at the Fermi level in the majority spin channel is much higher than that in the minority spin channel. Furthermore, the calculated magnetoresistance (MR) ratio of the MnCo^T-terminated interface reaches up to 2886%, while that of the MnMn^{T1}-terminated interface is only 330%. Therefore, Mn$_2$CoAl/Ag/Mn$_2$CoAl CPP-SV with an MnCo^T-terminated interface structure has a better application in a spintronics device.

Author Contributions: Conceptualization, Q.-Y.M.; Methodology, X.-Y.F.; Software, X.R.; Data Curation, Z.-Y.L.; Writing-Original Draft Preparation, Z.C. and X.R.; Writing-Review & Editing, M.-S.W.; Supervision, Y.F.

Funding: This research was funded by [National Natural Science Foundation of China] grant number [11747114], and by [Doctor Foundation of Jiangsu Normal University] grant number [17XLR046 and 16XLR022], and by [Xuzhou Technology Plan Project] grant number [KC16SQ179].

Conflicts of Interest: The author declares no conflict of interest.

References

1. Wolf, S.A.; Awschalom, D.D.; Buhrman, R.A.; Daughton, J.M.; von Molnár, S.; Roukes, M.L.; Chtchelkanova, A.Y.; Treger, D.M. Spintronics: A spin-based electronics vision for the future. *Science* **2001**, *294*, 1488–1495. [CrossRef] [PubMed]

2. Li, X.; Yang, J. First-principles design of spintronics materials. *Natl. Sci. Rev.* **2016**, *3*, 365–381. [CrossRef]

3. Nakatani, T.M.; Furubayashi, T.; Kasai, S.; Sukegawa, H.; Takahashi, Y.K.; Mitani, S.; Hono, K. Bulk and interfacial scatterings in current-perpendicular-to-plane giant magnetoresistance with $Co_2Fe(Al_{0.5}Si_{0.5})$ Heusler alloy layers and Ag spacer. *Appl. Phys. Lett.* **2010**, *96*, 212501. [CrossRef]

4. Childress, J.R.; Carey, M.J.; Cyrille, M.C.; Carey, K.; Smith, N.; Katine, J.A.; Boone, T.D.; Driskill-Smith, A.A.G.; Maat, S.; Mackay, K.; et al. Fabrication and recording study of all-metal dual-spin-valve CPP read heads. *IEEE Trans. Magn.* **2006**, *42*, 2444–2446. [CrossRef]

5. Graf, T.; Felser, C.; Parkin, S.S.P. Simple rules for the understanding of Heusler compounds. *Prog. Solid State Chem.* **2011**, *39*, 1–50. [CrossRef]

6. Felser, C.; Fecher, G.H. (Eds.) *Spintronics*; Springer: Berlin, Germany, 2013.

7. Galanakis, I.; Dederichs, P. (Eds.) Half-metallicity and Slater-Pauling behavior in the ferromagnetic Heusler alloys. In *Half-Metallic Alloys*; Springer: Berlin, Germany, 2005.

8. Feng, Y.; Chen, H.; Yuan, H.K.; Zhou, Y.; Chen, X. The effect of disorder on electronic and magnetic properties of quaternary Heusler alloy CoFeMnSi with LiMgPbSb-type structure. *J. Magn. Magn. Mater.* **2015**, *378*, 7–15. [CrossRef]

9. Wang, X.T.; Cheng, Z.; Liu, G.; Dai, X.; Rabah, K.; Wang, L.; Bouhemadou, A. Rare earth-based quaternary Heusler compounds MCoVZ (M = Lu, Y; Z = Si, Ge) with tunable band characteristics for potential spintroic applications. *IUCrJ* **2017**, *4*, 758–768. [CrossRef] [PubMed]

10. Jourdan, M.; Minár, J.; Braum, J.; Kronenerg, A.; Chadov, S.; Balke, B.; Gloskovskii, A.; Kolbe, M.; Elmers, H.J.; Schönhense, G.; et al. Direct observation of half-metallicity in the Heusler compound Co_2MnSi. *Nat. Commun.* **2014**, *5*, 3974. [CrossRef] [PubMed]

11. Galanakis, I.; Dederichs, P.H.; Papanikolaou, N. Slater-Pauling behavior and origin of the half-metallicity of the full-Heusler alloys. *Phys. Rev. B* **2002**, *66*, 174429. [CrossRef]

12. Takahashi, Y.K.; Hase, N.; Kodzuka, M.; Itoh, A.; Koganezawa, T.; Furubayashi, T.; Li, S.; Varaprasad, B.S.D.C.S.; Ohkubo, T.; Hono, K. Structure and magnetoresistance of current-perpendicular-to-plane pseudo spin valves using $Co_2Mn(Ga_{0.25}Ge_{0.75})$ Heusler alloy. *J. Appl. Phys.* **2013**, *113*, 223901. [CrossRef]

13. Sakuraba, Y.; Ueda, M.; Miura, Y.; Sato, K.; Bosu, S.; Saito, K.; Shirai, M.; Konno, T.; Takanashi, K. Extensive study of giant magnetoresistance properties in half-metallic Co_2(Fe,Mn)Si-based devices. *J. Appl. Phys. Lett.* **2012**, *101*, 252408. [CrossRef]

14. Sakuraba, Y.; Izumi, K.; Miura, Y.; Futasukawa, K.; Iwase, T.; Bosu, S.; Saito, K.; Abe, K.; Shirai, M.; Takanashi, K. Mechanism of large magnetoresistance in $Co_2MnSi/Ag/Co_2MnSi$ devices with current perpendicular to the plane. *Phys. Rev. B* **2010**, *82*, 094444. [CrossRef]

15. Li, S.; Takahashi, Y.K.; Furubayashi, T.; Hono, K. Enhancement of giant magnetoresistance by $L2_1$ ordering in $Co_2Fe(Ge_{0.5}Ga_{0.5})$ Heusler alloy current-perpendicular-to-plane pseudo spin valves. *Appl. Phys. Lett.* **2013**, *103*, 042405. [CrossRef]

16. Skaftouros, S.; Özdoğan, K.; Şaşioğlu, E.; Galanakis, I. Generalized Slater-Pauling rule for the inverse Heusler compounds. *Phys. Rev. B* **2013**, *87*, 024420. [CrossRef]

17. Feng, L.; Tang, C.; Wang, S.; He, W. Half-metallic full-Heusler compound Ti_2NiAl: A first-principles study. *J. Alloy. Compd.* **2011**, *509*, 5187–5189.

18. Feng, Y.; Wu, B.; Yuan, H.K.; Kuang, A.L.; Chen, H. Magnetism and half-metallicity in bulk and (1 0 0) surface of Heusler alloy Ti_2CoAl with Hg_2CuTi-type structure. *J. Alloy. Compd.* **2013**, *557*, 202–208. [CrossRef]

19. Jakobsson, A.; Mavropoulos, P.; Şaşioğlu, E.; Blügel, S.; Ležaić, M.; Sanyal, B.; Galanakis, I. First-principles calculations of exchange interactions, spin waves, and temperature dependence of magnetization in inverse-Heusler-based spin gapless semiconductors. *Phys. Rev. B* **2015**, *91*, 174439. [CrossRef]

20. Ouardi, S.; Fecher, G.H.; Felser, C.; Kübler, J. Realization of Spin Gapless Semiconductors: The Heusler Compound Mn_2CoAl. *Phys. Rev. Lett.* **2013**, *110*, 100401. [CrossRef] [PubMed]

21. Galanakis, I.; Özdoğan, K.; Şaşioğlu, E.; Blügel, S. Conditions for spin-gapless semiconducting behavior in Mn_2CoAl inverse Heusler compound. *J. Appl. Phys.* **2014**, *115*, 093908. [CrossRef]

22. Jamer, M.E.; Assaf, B.A.; Devakul, T. Magnetic and transport properties of Mn$_2$CoAl oriented films. *Appl. Phys. Lett.* **2013**, *103*, 142403. [CrossRef]

23. Xu, G.Z.; Du, Y.; Zhang, X.M.; Zhang, H.G.; Liu, E.K.; Wang, W.H.; Wu, G.H. Magneto-transport properties of oriented Mn$_2$CoAl films sputtered on thermally oxidized Si substrates. *Appl. Phys. Lett.* **2014**, *104*, 242408. [CrossRef]

24. Skaftouros, S.; Özdoğan, K.; Şaşioğlu, E.; Galanakis, I. Search for spin gapless semiconductors: The case of inverse Heusler compounds. *Appl. Phys. Lett.* **2013**, *102*, 022402. [CrossRef]

25. Wang, X.L. Proposal for a new class of materials: Spin gapless semiconductors. *Phys. Rev. Lett.* **2008**, *100*, 156404. [CrossRef] [PubMed]

26. Feng, Y.; Zhou, T.; Chen, X.; Yuan, H.K.; Chen, H. Thermodynamic stability, magnetism and half metallicity of Mn$_2$CoAl/GaAs (0 0 1) interface. *J. Phys. D Appl. Phys.* **2015**, *48*, 285302. [CrossRef]

27. Zarei, S.; Hashemifar, S.J.; Akbarzadeh, H.; Hafari, H.J. Half-metallicity at the Heusler alloy Co$_2$Cr$_{0.5}$Fe$_{0.5}$Al (001) surface and its interface with GaAs (001). *Phys. Condens. Matter* **2009**, *21*, 055002. [CrossRef] [PubMed]

28. Perdew, J.P.; Burke, K.; Ernzerhof, M. Generalized gradient approximation made simple. *Phys. Rev. Lett.* **1996**, *77*, 3865–3868. [CrossRef] [PubMed]

29. Blöchl, P.E. Projector augmented-wave method. *Phys. Rev. B* **1994**, *50*, 17953. [CrossRef]

30. Ghader, N.; Hashemifar, S.J.; Akbarzadeh, H.; Peressi, M. First principle study of Co$_2$MnSi/GaAs(001) heterostructures. *J. Appl. Phys.* **2007**, *102*, 074306. [CrossRef]

31. Taylor, J.; Guo, H.; Wang, J. Ab initio modeling of quantum transport properties of molecular electronic devices. *Phys. Rev. B* **2001**, *63*, 245407. [CrossRef]

32. Waldron, D.; Haney, P.; Larade, B.; MacDonald, A.; Guo, H. Nonlinear spin current and magnetoresistance of molecular tunnel junctions. *Phys. Rev. Lett.* **2006**, *96*, 166804. [CrossRef] [PubMed]

33. Burrows, C.W.; Bastiman, S.A.; Bell, G.R. Interaction of Mn with GaAs and InSb: Incorporation, surface reconstruction and nano-cluster formation. *J. Phys. Condens. Matter* **2014**, *26*, 395006. [CrossRef] [PubMed]

34. Feng, Y.; Chen, X.; Zhou, T.; Yuan, H.K.; Chen, H. Structural stability, half-metallicity and magnetism of the CoFeMnSi/GaAs (0 0 1) interface. *Appl. Surf. Sci.* **2015**, *346*, 1–10. [CrossRef]

35. Van de Walle, C.G.; Neugebauer, J. First-principles calculations for defects and impurities: Applications to III-nitrides. *J. Appl. Phys.* **2004**, *95*, 3851. [CrossRef]

MDPI

St. Alban-Anlage 66

4052 Basel

Switzerland

Tel. +41 61 683 77 34

Fax +41 61 302 89 18

www.mdpi.com

Applied Sciences Editorial Office

E-mail: applsci@mdpi.com

www.mdpi.com/journal/applsci